The Selectivity of Drugs

Other titles in the series

OUTLINE STUDIES IN BIOLOGY

General Editor : Professor J.M. Ashworth, University of Essex

Editors' Foreword

The student of biological science in his final years as an undergraduate and his first years as a postgraduate is expected to gain some familiarity with current research at the frontiers of his discipline. New research work is published in a perplexing diversity of publications and is inevitably concerned with the minutiae of the subject. The sheer number of research journals and papers also causes confusion and difficulties of assimilation. Review articles usually presuppose a background knowledge of the field and are inevitably rather restricted in scope. There is thus the need for short but authoritative introductions to those areas of modern biological research which are either not dealt with in standard introductory textbooks or are not dealt with in sufficient detail to enable the student to go on from them to read scholarly reviews with profit. This series of books is designed to satisfy this need.

The authors have been asked to produce a brief outline of their subject assuming that their readers will have read and remembered much of a standard introductory textbook of biology. This outline then sets out to provide by building on this basis, the conceptual framework within which modern research work is progressing and aims to give the reader an indication of the problems, both conceptual and practical, which must be overcome if progress is to be maintained. We hope that students will go on to read the more detailed reviews and articles to which reference is made with a greater insight and understanding of how they fit into the overall scheme of modern research effort and may thus be helped to choose where to make their own contribution to this effort.

These books are guidebooks, not textbooks. Modern research pays scant regard for the academic divisions into which biological teaching and introductory textbooks must, to a certain extent, be divided. We have thus concentrated in this series on providing guides to those areas which fall between, or which involve, several different academic disciplines. It is here that the gap between the textbook and the research paper is widest and where the need for guidance is greatest. In so doing we hope to have extended or supplemented but not supplanted main texts, and to have given students assistance in seeing how modern biological research is progressing, while at the same time providing a foundation for self help in the achievement of successful examination results.

The Selectivity of Drugs

Adrien Albert

Professor Emeritus, Australian National University, Canberra

Chapman and Hall

London

First published in 1975
by Chapman and Hall Ltd
11 New Fetter Lane, London EC4P 4EE

Typeset by E.W.C. Wilkins Ltd., London and Northampton and
printed in Great Britain by William Clowes & Sons Ltd.,
London, Colchester and Beccles

ISBN 0 412 13090 4

Distributed in the USA
by Halsted Press, a Division
of John Wiley & Sons, Inc. New York

Library of Congress Catalog Card Number 74–22170

Contents

1 Introduction

1.1 What is Selectivity?

A biologically-active substance is said to be selective if it strongly affects certain cells without causing any change in others, even when the two kinds of cells are close neighbours. In living organisms, there are many substances, often quite small molecules, which have been chosen for their specificity. This choice has been made under the strong pressure of natural selection, unhurried by any consideration of time. Such chemical compounds operate the metabolism of the cells and tissues, and ensure their health, survival, and reproduction. Important among the smaller of these selective molecules are vitamins, coenzymes, hormones, neurotransmitters, inorganic ions, nutritional fragments, respiratory and photosynthesizing pigments; also adenosine triphosphate (ATP), which is the energy store of every living creature; and the purine and pyrimidine components of deoxyribonucleic acid (DNA) on which is encoded the genetic information that guides and controls each organism. The interaction of each of these small molecules with its complementary biopolymer (most often an enzyme) generates the needed physiological response and affords a typical example of what is becoming known as cellular recognition [1].

Different from these essential cell constituents, although often related to them, are the biologically-active agents used for the treatment of disease in man and his economic animals, and in agriculture to keep crops free from pests. The word *drug* denotes a substance used for the first purpose, and *crop protective agent* for the second, but there is no fundamental difference of principle in their mode of action.

Drug therapy has two, fundamentally opposed divisions. The first of these strives to improve the action of one of the cell's natural agents by modifying the molecule in order to localize or intensify its action. For instance, the solubility can be decreased to make it form a deposit, or a change is made so that it becomes a poorer fit on the naturally-occurring destructive enzyme. Both of these devices have proved useful in therapy, e.g. with steroid hormones. Such drugs, which seek to improve on Nature by performing more desirably, are called *agonists.*

However, most drugs are not agonists but *antagonists,* devised to counteract a naturally-arising effect when this, as in illness, becomes undesirable. Such drugs are designed to overpower an invading species (a process called chemotherapy) or to help a disordered metabolism by suppressing abnormal or unwanted cycles or rhythms (pharmacodynamic therapy). The achievement of these results requires application of what is called *selective toxicity.* Typical of antagonists are all the general anaesthetics, local anaesthetics, hypnotics, antihistaminics, analgesics, anti-epileptics, and antagonists of such neuro-transmitters as acetylcholine, noradrenaline, and serotonin. Other selective toxicants are used to rid man and his domestic animals of invading organisms, whether bacteria, viruses, fungi, protozoa, or worms; or to defend crops against insects, fungi, or weeds.

Most of these selective agents are organic chemicals of low molecular weight, synthesized in the chemical laboratory. However, in some instances they are naturally-occurring substances used far away from their natural context, e.g. alkaloids (generally considered to be the nitrogenous excreta of plants), antibiotics (waste substances produced by bacteria or fungi, after they have completed a phase of fast growth, seemingly to inactivate their own growth [2]), and unphysiologically large doses of hormones (e.g. cortisone for arthritis) or vitamins. Immunochemicals (the antigens and antibodies of vaccine and sera respectively) are separated from the subject of this book by their huge molecular weight.

It is convenient to refer to the cells which an antagonist is required to spare as the *economic species,* and those which the antagonist must influence as the *uneconomic species.*

The concept of selective toxicity is admirably illustrated by general anaesthetics. The more toxic the anaesthetic, the greater its value in medicine. Surprising as this statement may seem, it is true; but the toxicity must be (i) highly selective for the central nervous system, (ii) give a gently graded response, and (iii) be completely reversible with time. In fact, from the introduction of ether in 1846 to the present day, all successful anaesthetics have combined a high toxicity for the central nervous system with negligible toxicity to other tissues, and all toxicity has rapidly and completely disappeared when administration ceased. These are the qualities desired throughout the whole territory of pharmacodynamic therapy, but in chemotherapy irreversible toxicants are preferred.

The selective eradication of tumours (whether benign or malignant) constitutes a special application of antagonists, because the territory lies precariously between pharmacodynamics and chemotherapy. In this area, although the economic and uneconomic cells are part of the same organism, it is highly desirable that the toxic action of the drug (preferably highly selective) should be permanent.

The subject of selective toxicity has tremendous scope and has been presented in considerable detail elsewhere [3]. An alternative to selective toxicity is biological control, usually by broadcasting predatory insects, bacteria, or fungi. Although some remarkable successes have been achieved in this area, their total number is few in spite of the large expenditure of effort. It seems likely that, for many years to come, the selective manipulation of uneconomic species will be effected mainly by chemical substances of high specificity.

1.2 The aims and accomplishments of Selectivity

A current reminder of what selective agents have accomplished, and what is still required of them, can be found in the *Annual Report of the Director-General to the World Health Assembly and to the United Nations,* and in two *World Health Organization* monthlies (the more general *Chronicle* and the rather specialized *Bulletin*). These are published in Geneva (Switzerland) and keep the health of the whole world under survey.

Infectious diseases. The disease which, throughout the world, causes the greatest amount of debility and illness (and death if untreated) is malaria. The World Health Organization (WHO) gives top priority to the elimination of this disease by advising on draining, spraying, and medication (both prophylactic and curative), and by constant experimentation in the field to improve these methods. Projects submitted by an undeveloped or developing nation and approved by WHO usually obtain international funding. Some 720 million people, mainly in the tropics, living in areas that were highly malaria-prone in the 1940's are now without risk of infection. Another 630 million are kept malaria-free by sprays and prophylactic drugs. Unfortunately, 480 million others are living where malaria is prevalent (over 200 million of them in Africa). Moreover substantial pockets are developing where the causative organisms, or the mosquitos that convey them, have become resistant to the most used toxicants. Here lies a great challenge to discover new types of selective agents.

After malaria, trypanosomiasis, leishmaniasis, and amoebiasis are the most serious of the diseases caused by protozoa. Excellent drugs exist, but the emergence of resistant strains of trypanosomes points to the constant need to discover new drugs, particularly ones acting by different mechanisms. In these, as in other diseases common in the poorer nations, improvement in nutrition, sanitation, and water supply are often required to supplement the beneficial effects of selective agents.

Diseases caused by bacteria have been mastered by the developed countries only since the late 1930's. Until then, the medical wards of even the world's most renowned hospitals always had many patients severely ill with, or dying of, pneumonia. Special hospitals, beyond the confines of the city, had to be maintained for tuberculosis, and in these countless patients wasted away and died. In the surgical wards, severe, disabling bacterial infections of the arms and legs were common but difficult to improve, bacterial infection of the bladder was widespread among elderly men, and peritonitis remained a dreaded complication of abdominal surgery for which little could be done. Mothers in childbirth were at special risk of septicaemia, often fatal. In the children's wards, osteomyelitis was intractible and severe middle ear infection, often leading to permanent deafness, abounded. Yet modern chemotherapy with such selective drugs as penicilin, isoniazid, sulphonamides, or tetracyclines has provided rapid cures for all such infections. Bacterial epidemics, too, are now far less dreaded.

However there are seven bacterial diseases still insufficiently controlled in the world (cholera, tuberculosis, trachoma, leprosy, brucellosis, gonorrhoea, and syphilis). The five former give most trouble in underdeveloped and developing countries, but the two latter are tending to increase most in developed countries. Effective drugs are available, but various social factors are hindering their application.

Of virus-caused diseases, smallpox is the most prevalent and damaging, although widespread vaccination and quarantine regulations are steadily decreasing the incidence. Only in recent years have clues been found for preventing and treating virus diseases with drugs, but this is a very important field because the current recourse to immunotherapy with sera and vaccines is successful for only a few other viral diseases (among them yellow fever, measles, and poliomyelitis).

Turning to the fungal diseases of man, these (sometimes even when superficial) are being treated by internal medication, but there is still scope for better drugs.

Of infections in man caused by parasitic worms, schistosomiasis (bilharziasis), which is snail-transmitted, is the worst; WHO estimates the world incidence as 200 million, steadily rising due to new irrigation projects in infected countries. Improved drugs for treatment are available, but a good prophylactic is needed. New molluscicides, acting against the infected snails, are helping the situation in Egypt. Necessary improvements in hygiene are hard to enforce in hot countries if the population is large relative to the available water supply. Filariasis, a tropical mosquito-borne worm infection, responds well to medication with diethylcarbamazine combined with spraying against larvae, but there are still 100 million sufferers, some with elephantiasis. Hookworm, which penetrates the skin of about 500 million agricultural workers in the tropics and causes debilitation, responds well to modern drugs. Roundworms, common in the tropics with about 650 million sufferers, are easily killed with the new synthetic anthelmintics, as are the universally occurring threadworms and tapeworms.

Many farm animals suffer from severe worm diseases which sap their vitality and decrease their market value. Effective anthelmintics are known but many of the best are too expensive to be economic.

Non-infectious diseases. Whereas in underdeveloped countries most cases of illness stem from infectious diseases, most of the illness in the more prosperous countries is metabolic and hence requires pharmacodynamic rather than chemotherapeutic agents. In these developed countries, the principal causes of death are (in decreasing order of frequency) heart disease, cancer, and vascular lesions affecting the central nervous system (causing strokes). These account for 70 per cent of all deaths. In addition, mental ill-health and rheumatoid diseases account for a high percentage of incapacitation. For each of these diseases some drugs are available, but more effective ones are urgently required. The last 20 years has seen severe mental illness yield far more to medication than to psychological treatment; consequently much research on biochemical causes of insanity is under way.

Cancer, a collective term for over 100 types of malignant tumour, is yielding more and more to medication so that it is now possible to speak of cure by selectively toxic drugs. Unfortunately for only a few of the very common slow-growing solid tumours have remedies been found and some new leads are required here.

All these successes have not been achieved without some disadvantages. By the third quarter of this century, selective medication had so increased the survival rate, particularly in developing countries, that food shortages began to appear. This imbalance is being vigorously fought by the Food and Agricultural Organization, through its advice on better land management. In this endeavour, a significant contribution is being made by selectively toxic agents specially devised for agricultural use. As late as 1967 it was found that fungal diseases were wasting about 12 per cent of the world's annual crops, insects about 14, and weeds another 9 per cent, an estimated total loss of 1400 million tons to which must be added much spoilage by rodents.

In concluding this chapter, it must be emphasized that much more remains to be done with selective agents than has yet been accomplished, great as the achievements are. However, the successes have contributed so much to health and food resources that it is reasonable to expect that more of the outstanding problems could be solved in the next decade or two. Such a happy outcome will require that all known physicochemical properties controlling selectivity be kept under constant review [3], and the information applied to current research projects.

References

[1] Greaves, M.F. (1975), *Cellular Recognition*, Chapman and Hall, London, 64 pp.

[2] Woodruff, H. (1966), in *Biochemical Studies of Antimicrobial Drugs* (ed. Newton, B. and Reynolds, P.), University Press, Cambridge.

[3] Albert, A. (1973), *Selective Toxicity*, 5th Edn, Chapman and Hall, London, 597 pp.

2 Three principles that control selectivity

As far as is yet known, selective agents exert their favourable effects through one or more of the following three principles. *Either* they are accumulated principally by the uneconomic species, *or,* utilizing the differences recognized in comparative biochemistry, they injure a biochemical system that is important only for the uneconomic species, *or* they react exclusively with a cell structure that exists only in the uneconomic species. These three principles will be outlined in the three following paragraphs.

Selectivity through distribution permits use of an agent toxic to both economic and uneconomic cells, provided that it is accumulated only by the latter. In 1927, it was discovered in France that 10 per cent sulphuric acid could be safely sprayed on cereal crops to destroy the weeds, and this result has repeatedly been confirmed. Sulphuric acid is, of course, injurious to the cytoplasm of both wheat and weed, but it never reaches that of the former for two reasons. The exterior of cereal grasses is smooth and waxy, whereas that of the usual weeds (which are dicotyledons) is rough and wax-free; hence the acid runs off the former but is accumulated by the latter. Moreover the tender new shoots of cereals are protected by leaf-sheaths, whereas the growing point of dicotyledons is exposed and vulnerable [1]. This principle is expanded in Chapter 4.

Selectivity through comparative biochemistry. The most striking biochemical differences between cells have most often been observed in the processes of synthesis, less often in degradation, and seldom in energy storage. The high selectivity of the sulphonamide antibacterials depends on the fact that pathogenic bacteria cannot absorb folic acid or its derivatives, but can synthesize them from *p*-aminobenzoic acid, a process blocked by the sulphonamides. Mammals on the other hand cannot synthesize folic acids (and hence they tolerate sulphonamides well), but they do not lack folic acid which they easily absorb from food. This kind of selectivity is covered in more detail in Chapter 5.

Selectivity through comparative cytology. Outstanding cytological differences between plants and animals have long been recognized. Thus cell walls, and photosynthetic apparatus, are found in plants but not in animals; likewise nerve and muscle cells are found in animals, but not in plants. The electron microscope has shown that the cell itself is full of component parts, and that each kind of organelle, as these parts are called, displays special differences. Moreover, there are differences between cell components from different tissues in the same organism. Penicillin and chloramphenicol are selective because they interfere with structures found only in bacteria. This topic will be developed further in Chapter 6.

Reference

[1] Blackman, G. (1946), *Agriculture,* **53**, 16–24.

3 Steps in the correlation of structure with biological activity

3.1 Introduction

The previous chapter has outlined the sources of selectivity which are available for controlling uneconomic cells. Before these principles are discussed in depth, a brief account must be given of what is known of the molecular basis for biological activity in a foreign substance. This activity can usefully be viewed as a primary force which knowledge of selectivity can tame in the service of man.

It is now recognized that, in organisms, the characteristic biological response of each enzyme and hormone is highly dependent on fine details of its chemical structure. For instance, the activity of thiamine (vitamin B_1) drops to 5 per cent if the methyl group is removed from the pyrimidine ring and to less than 1 per cent if the methyl group in the thiazole ring is deleted. Moreover, activity completely disappears if an extra methyl group is inserted [1]. However side-chains can often be modified without harm to the biological effect; thus removal of the long hydrocarbon side-chain of vitamin K_1 affects its action very little.

Synthetic drugs usually depend similarly on minute details of structure. Consider, for example, the molecule of benzenesulphonamide (*3.1*) which has three different positions into which an amino group ($-NH_2$) can be inserted. In two of these it gives rise to an inactive substance, but in the third, (the para position), it produces the highly antibacterial substance sulphanilamide. Again, in acridine (*3.2*), an amino group can be inserted in five different positions: in three of these it gives amino-acridines that are almost inactive, but in the 3- and the 9-positions it produces powerful antibacterials. Still more strikingly, in quinoline (*3.3*), a hydroxy group ($-OH$) can be inserted in seven different positions: in six of these it gives completely inert substances, but in the remaining position (no. 8) a strongly anti-bacterial and antifungal substance is produced. The reasons why the active isomers are active, the the inactive ones inactive, are well understood and will be set out in what follows. However, before reviewing these details, it should help to stand back a little and take a broader view of relationships between structure and activity.

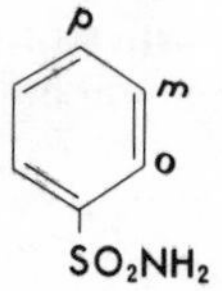

p m o SO₂NH₂

Benzenesulphonamide
(*3.1*)

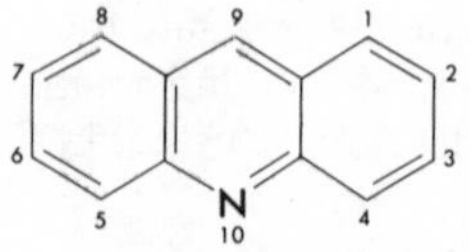

8 9 1 7 2 6 3 5 N 10 4

Acridine
(*3.2*)

5 4 6 3 7 2 8 N 1

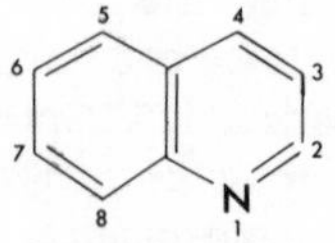

Quinoline
(*3.3*)

3.2 The first correlations

By the middle of the last century, long before Arrhenius introduced his idea of electrolytic dissociation, it was known that the biological activity of a salt was due to its basic *or* its acidic component and not to the whole salt. Thus the poisonous entity in lead acetate and lead nitrate was recognized as the lead moiety and it was clearly understood that the toxicity of sodium, potassium, and calcium arsenites resided only in the arsenite portion of these salts [2].

The first correlation of structure with activity in organic chemicals was made in 1869 when Crum Brown and Fraser, in Scotland, showed that several alkaloids, when quaternized, lost their characteristic pharmacological properties and acquired the muscle-relaxing powers of tubocurarine, itself a quaternary amine whose site of action was known to lie at the junction of nerve and muscle. Thus when strychnine, on heating with methyl iodide, became quaternized by methylation on the ring nitrogen atom, its properties changed from a convulsant to a curariform relaxant [3]. No other examples came to light of a single chemical group being able to confer a single pharmacological action on a variety of complicated nuclei. A solution of this puzzling correlation came only in the present century, for it had to await the postulation of receptors for drugs and analogous phenomena in enzyme chemistry.

3.3 The concept of 'receptors'

Three striking characteristics of the action of drugs indicate very strongly that they are concentrated by cells on small, specific areas named receptors by Ehrlich [4]. These properties are, (i) the high dilution (e.g. 10^{-9} M) at which solutions of many drugs retain their activity, (ii) the high chemical specificity of drugs, so discriminating that even the D- and L-isomers of a single substance can have different pharmacological action, and (iii) the high biological specificity of drugs, e.g. the powerful effect that adrenaline exerts on cardiac, but not on striated muscle. The idea of receptors became more firmly established by the quantitative work of A.J. Clark (begun about 1920) who showed that drug-receptor combinations obeyed the law of mass action [5]. He further pointed out that most of the quantitative data in the literature could be interpreted as resulting from the formation of a reversible bond between a drug and its specific receptor.

3.4 The receptor as an enzyme or permease

Even in the earliest discussions on receptors, it was noted that the reaction of an enzyme with its coenzyme and antagonist had much in common with the reaction of a receptor with an agonist or an antagonist, respectively. The great specificity of structure (often even stereospecificity) served to increase a feeling that these were parallel phenomena.

By 1910, too, it was well known that many enzymes could be blocked by substances similar in molecular structure to the normal substrates. Although suspected, no close connexion between this discovery and the action of drugs was established until 1926 when Loewi and Navatril showed that the alkaloid physostigmine caused the heart to contract, not by any direct action, but by blocking the local enzyme, acetylcholinesterase. This blockade clearly allowed environmental acetylcholine to act on the heart, and thus enzymes were established as one kind of receptor [6].

It was soon discovered that the portion of the physostigmine molecule that inhibited esterases was a type of urethane, namely the methylcarbamoyloxy group (*3.4*). When this, or even the carbamoyloxy group, was inserted into much simpler basic nuclei, potent esterase inhibitors were obtained. The much used drug carbachol (*3.5*) was obtained through this process of simplification.

By 1934 it was agreed that acetylcholine

$-O\cdot\overset{O}{C}\cdot NHMe$ (3.4)

$Me_3\overset{+}{N}\cdot CH_2\cdot CH_2-O\cdot\overset{O}{C}\cdot NH_2$
Carbachol (cation)
(3.5)

$Me_3\overset{+}{N}\cdot CH_2\cdot CH_2\cdot O\cdot\overset{O}{C}\cdot Me$
Acetylcholine (cation)
(3.6)

(*3.6*) is the natural substance that transmits all nervous impulses between nerve and muscle [7]. Ing quickly saw that tubocurarine (and a great many other quaternary amines) blocked neuromuscular transmission by competing with acetylcholine for the latter's receptors which these amines could block but not operate [8]. It was now clear for the first time that drugs with a given group (in this case the quaternary ammonium group) could exert two quite opposite actions. They could have an *agonist*-type action if they could mimic a natural metabolite, at least well enough to prevent its destruction, and carbachol acts in this way. However they could have an *antagonist*-type action if they differed from the metabolite enough to interfere with its natural uptake or use. It is now known that the acetylcholine receptor is not an enzyme, like acetylcholinesterase, but a permease i.e. a membrane-situated regulator of the permeability of inorganic ions. Recently isolated, it was found to be a very hydrophobic protein [9].

These relationships were not extended from pharmacodynamics to chemotherapy until Woods (1940) demonstrated the reversal of the antibacterial action of sulphanilamide (*3.7*) by *p*-aminobenzoic acid (*3.8*), pointing out that this reversal depended on the structural similarity of the two substances [10]. Later the receptor for sulphonamides was found to be the enzyme, dihydrofolate synthetase, which incorporates *p*-aminobenzoic acid into the molecule of dihydrofolic acid (*5.3*), an essential coenzyme for the biosynthesis of purines and thymine, and hence of DNA. This enzyme was isolated and purified by Brown in 1962 [11].

It is noteworthy that it is not the mere presence of a sulphonamide group that enables certain benzenesulphonamides to be potent inhibitors of this enzyme and successful antibacterial drugs. In addition, they must have sufficient resemblance to *p*-aminobenzoic acid for the enzyme to be deceived into absorbing them. The essentials have been found to be: a primary amino-group, which must be situated *para* to the sulphonamide group, plus a similar polar charge and steric arrangement at the sulphonamide-surrogate end, and the same, critical distance must be maintained between these features in an absolutely flat molecule. A further requirement is that a molecule of these specifications should still be sufficiently unlike *p*-aminobenzoic acid, in minor details, not to function as a substitute for it. Many other useful drugs have been found which contain the sulphonamide group, e.g. some diuretics and most oral antidiabetics; but these do not meet the above specifications and consequently are not antibacterial. The antibacterials, on the other hand, can meet the above specifications, without containing sulphur!

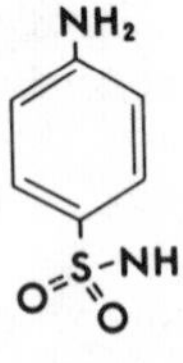

NH_2 / $O=S(=O)-NH_2$
Sulphanilamide
(3.7)

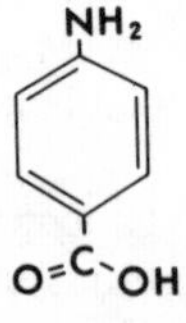

NH_2 / $O=C-OH$
p-Aminobenzoic acid
(3.8)

3.5 The receptor as a nucleic acid

This deep and practically complete understanding of the mode of action of sulphonamide antibacterials put an end to the older notion that a particular group or nucleus could introduce a characteristic type of biological action into a molecule. Instead the interests of drug designers became focused on (a) steric properties, which controlled access to the correct receptor and a good fit on arrival there, and (b) electron distribution, at first simply as recognition of areas of positive and negative polarization in the molecule.

Both of these factors came to the fore in studies of the aminoacridines (topical antibacterials much used in wounds in the Second World War). First it was shown that the biological action of aminoacridines was proportional to the fraction ionized as cation [12], a first step in uncovering the physical basis of their biological properties. Acridine itself (*3.2*) is only a weak base, with a pK_a of 5·3 (in water at 37°C), and hence only 1 per cent is ionized at pH 7. However two of the five monoaminoacridines are strong bases, and this strength (which ensures complete ionization) was traced to a resonance effect in their molecules not possible in the three other isomers [13]. This positive correlation between ionization* and bacteriostasis was demonstrated over a wide variety of bacterial species, anaerobes and aerobes, both Gram-positive and -negative [14] (see Table 3.1).

Table 3.1 Dependence of antibacterial action on ionization [13].

-acridine	*Minimal bacteriostatic concentration (Streptococcus pyogenes) 1 part in*	*Percentage ionized as cations (pH 7·3; 37°C)*
1-Amino-	10 000	2
2-Amino-	10 000	2
3-Amino-	80 000	73
4-Amino-	5 000	<1
9-Amino-	160 000	100
2, 7-Diamino-	20 000	3
3, 6-Diamino-	160 000	99
4, 5-Diamino-	<5 000	<1

*For more on ionization, see Section 4.4 (p. 29).

Many substituted acridines were then synthesized and tested, and it was always found that the substituents exerted no direct effect on the antibacterial action except in so far as they modified the ionization to a degree usually predictable from the sign and magnitude of the Hammett sigma constant of each substituent.

The acridine-type of antibacterial action has these characteristics: it takes place at high dilution, even in the presence of serum proteins, and without harm to mammalian tissues. Because no other cations known at that time had this combination of properties, it was tempting to attribute them to the presence of the acridine nucleus. Nevertheless, by bold stepwise alterations of the molecule, these properties were evoked in a wide range of aromatic nuclei (even non-heterocyclic ones). The important features for this kind of biological action were found to be: complete ionization as cation at pH 7, and a completely flat molecule with an area of not less than 39 sq. Å of flat surface [15]. Thus biological equivalents of potent aminoacridines were found in the phenanthridine and benzoquinoline series, and in the pyridines and quinolines if provided with big enough coplanar substituents, and even in guanidino-anthracenes.

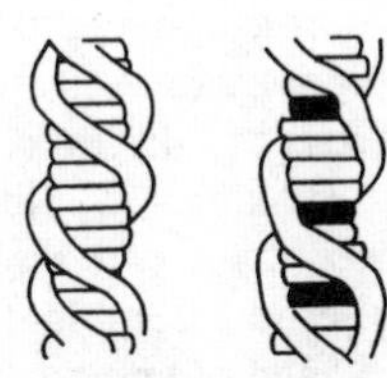

Fig. 3.1 Sketches representing the secondary structure of normal DNA (left) and DNA with intercalated molecules of proflavine (2, 8-diaminoacridine). The base-pairs and proflavine appear in edgwise projection, and the phosphate-deoxyribose backbone as a smooth coil [16].

That aminoacridines were accumulated by the nucleic acids of cells became clear through their use in vital staining [17]. The reason for the requirement for molecular flatness became evident in 1961 when Lerman showed that aminoacridine molecules were intercalated into DNA by stacking between layers of base-pairs, to which they were attached by van der Waals forces supplemented by stronger ionic bonds to the phosphate anions [18]. Figure 3.1 presents this situation diagramatically. The resultant increase in melting temperature showed that intercalation interfered with the unwinding of the DNA strands and hence with normal functioning.

In the very next year, it was demonstrated that aminoacridines injured bacteria by blocking the DNA template required by the enzymes that synthesise DNA and RNA [19].

These studies of structure-relationships in the aminoacridines established that nucleic acids, too, can be receptors. In fact, the drug-receptor interaction was observed here in unusual detail, much of it at the level of molecular biology. Many steroid hormones act by de-repressing a length of temporarily inactive DNA so that it can produce a messenger-RNA characteristic of that hormone.

3.6 The receptor as a coenzyme, or other small molecule

Yet another kind of drug-receptor relationship was brought to light by investigating the antimicrobial properties of 8-hydroxyquinoline, known for short as oxine [20]. Prior to this, it had been supposed that oxine owed its antibacterial properties to a combination of those of quinoline (*3.3*) and phenol in the one molecule. Yet neither quinoline nor phenol is at all antibacterial at a dilution of 1 in 5000, whereas oxine is active at 2 parts per million. That the biological properties of two substances could be combined by introducing their individual groups into a single molecule, strikes us today as absurd, because the favourable distribution of electrons in each component molecule must, far more often than not, be incompatible in the hybrid.

In the new study [20] it was found that the six isomers of oxine, obtained by moving the hydroxy-group to each other possible position (*3.3*), were non-antibacterial. This indicated that the biological properties of oxine were linked to its ability to chelate (i.e. to bind metal cations tightly by two or more well-separated atoms, forming a 5- or 6- membered ring). When tested, the six inert isomers failed to chelate whereas oxine showed strong chelation (and has long been used in the analysis of metals, for this property).

To sum up, the outstanding chelating and antibacterial properties of oxine are attributed to the juxtaposition of the oxygen and nitrogen atoms which permits the tight binding of heavy metal cations in a five membered ring, as shown in (*3.9*). The remaining positive charge on the metal can be removed by further addition of oxine, and the complexes become more liposoluble due to removal of the charge.

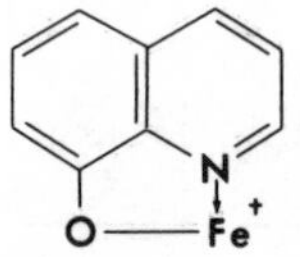

Oxine: 1 : 1–Fe complex (*3.9*)

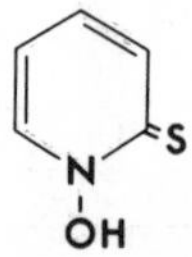

Pyrithione (*3.10*)

The question was then posed: does oxine act on bacteria by removing metals essential to bacterial welfare, or does it cause traces of metals to become more toxic to the bacteria? The latter proved to be the case, as first indicated by the following example of 'concentration quenching'. Staphylococci were completely killed in an hour by 0·00001 M oxine but were unharmed by 0.0007 M oxine; in fact even a saturated (0·005 M) solution would not kill them [21]. Streptococci behaved similarly. The meaning of this phenomenon became clear when it was found to occur only in media containing traces of iron or copper. The viability of staphylococci for 24 hours in distilled water permitted the decisive experiments, summarized in Table 3.2, to be made.

Table 3.2 The innocuousness of oxine in the absence of iron (bactericidal test) [20]. *Staphylococcus aureus*: pH 6–7 (20°)

Oxine mM	$FeSO_4$ *mM*	*Growth on plating our after 1 hr.*	
		In glass-distilled water	*In untreated meat broth*
nil	nil	prolific	prolific
0·01	nil	prolific	nil
nil	0·01	prolific	prolific
0·01	0·01	nil	nil

It is easy to see from Table 3.2 that oxine (0·00001 M) is biologically inert, but becomes bactericidal in the presence of a similar quantity of iron. Clearly the toxic agent is neither oxine nor iron, but the oxine–iron complex. When broth replaced water, no added iron was necessary because it was present in the medium. When the concentration of oxine was increased to 0·0013 M, the bactericidal action disappeared. This was attributed to formation of a non-antibacterial 2 : 1-oxine-iron complex because, when sufficient extra iron was added to the broth so that the 1 : 1 complex was re-formed, full bactericidal properties were restored [21].

In the absence of external heavy metals, oxine has been found to enter the cells of bacteria and fungi without harming them [22]. Yet if iron (for bacteria) or copper (for fungi) is present, the combination is rapidly lethal.

Other chelating antimicrobials have been found that, while having a totally different structure, mimic the action of oxine by being active only in the presence of a variable-valence metal, and hence show concentration quenching. Such a substance is 1-hydroxypyridine-2-thione (pyrithione) (*3.10*) [23], much used in the dermatology of the scalp. Another example is dimethyldithiocarbamic acid (*3.11*), whose salts are widely used as selective fungicides in agriculture. It is thought that all these metal complexes act by oxidatively destroying lipoic acid (thioctic acid) (*3.12*) which is the essential coenzyme for the oxidative decarboxylation of pyruvic acid, and accumulation of pyruvic acid has been demonstrated [24]. Here the receptor is apparently the small molecule (*3.12*). Another example of a coenzyme as receptor is furnished by the lethal action, in mammals, of hydrogen cyanide which follows directly from the binding of this poison to the free valence of iron in the porphyrin molecule of cytochrome oxidase.

$$Me_2N \cdot \overset{S}{\overset{\|}{C}} \cdot SH \qquad \overset{SH}{\overset{|}{CH_2}} \cdot CH_2 \cdot \overset{SH}{\overset{|}{CH}}(CH_2)_4 \cdot CO_2H$$

Dimethyldithiocarbamic acid (*3.11*) Lipoic acid (*3.12*)

3.7 Other aspects of receptors

The better understanding of receptors is bearing fruit in practical therapy. The existence of two different kinds of adrenergic receptor (α- and β-) was first suggested in 1948 by Ahlquist, and it was soon realized that the inhalation treatment of asthma depended on blocking the β-receptors in the lungs with isoprenaline. A more recent subdivision of β-receptors [25] led to the discovery that isoprenaline activates both β_2- receptors in the lung (desirable in therapy) and β_1-receptors in the heart (causing undesirable tachycardia). This differentiation led to selective screening procedures which furnished drugs, e.g. salbutamol, which activate only β_2- receptors [26].

In a few cases the union between agent and receptor is formed by covalent bonds. When this occurs, and the examples although few are important, the product is irreversible (or, at least, needs an unusual and powerful antidote). For example, the organic phosphate insecticides act by uniting covalently with the hydroxy group of serine in the enzyme acetylcholinesterase. Again, penicillin acts by inactivating the enzyme glycopeptide transpeptidase in the cytoplasmic membrane of bacteria by the formation of a covalent bond [27].

The action of a drug on a drug-receptor is often connected to the desired physiological result only through a long chain of other reactions, chemical and biological. Yet it is only with reference to the activity at some particular receptor that constitution-action studies have any validity. Thus a desired medicinal response can be sought in chemical structures quite remote from all that have so far furnished this response, because there is always the possibility of striking elsewhere in the chain of command.

3.8 Targets that are not necessarily receptors

Several chelating agents are routinely used in hospitals as antidotes for poisoning by inorganic substances. These antidotes circulate in the blood-stream without greatly depleting the body's essential metals, a tribute to the strength with which such cations are bound (usually by enzymes which they activate). Examples are dimercaprol (dithioglycerol), used to treat poisoning by compounds of arsenic, antimony, mercury, and gold; penicillamine, used for copper poisoning, particularly in Wilson's disease; and calcium ethylenediaminetetracetic acid (EDTA), a most effective remedy for lead poisoning. These antidotes have been provided with hydrophilic groups so that the chelated complexes cannot penetrate from the blood stream into cells and are easily excreted by the kidneys.

At the turn of the present century, Overton and Meyer independently put forward the lipoid theory of narcosis. This stated that chemically inert substances, of widely different molecular structures, exert depressant properties on those cells that are particularly rich in lipids; and that the higher the partition coefficient of the agent (between any lipoidal substance and water) the greater the depressant action. If after the words 'partition coefficient' we are allowed to insert 'up to the point where hydrophilic properties are almost lost', this statement is as true today as when it was written. Corwin Hansch pointed out later that the relationship is parabolic [28], because substances which are too lipophilic become trapped in external lipids and do not enter the cell.

Depressants are substantially non-ionized substances. They may be hydrocarbons, halogenated hydrocarbons, alcohols, ethers, ketones, weak acids like the barbiturates, weak bases, or aliphatic sulphones. These are the substances with hypnotic and general anaesthetic action; in higher concentrations they are used as volatile insecticides, particularly in the fumigation of stored grain. Because the spinal cord and brain are clad in lipid-rich membranes, the depressants become segregated there and they influence the central nervous system selectively through their high local concentration. They seem to interrupt nerve communication (reversibly, fortunately) simply by being foreign matter. This is the only kind of biological activity in which structure simply does not matter.
(For further discussion of partition coefficients and relative lipophilicity, see Section 4.4).

3.9 Regression analysis

Interest in partition coefficients extends far beyond the depressants, for this reason: favourable *balance* between hydrophilic and lipophilic properties could help any type of drug to reach its receptor and, although only a secondary property, cannot be neglected.

For depressant drugs, Hansch and his colleagues have devised the following regression equation, in which the squared term ensures a parabolic relationship:

$$\log(1/C) = k(\log P) - k'(\log P)^2 + k''$$

where P is the partition coefficient, C the concentration that produces a standard biological response, and k, k', and k'' constants, chosen with the help of the method of least squares, to make a 'best fit' of the experimental results. Because r (the statistician's correlation coefficient) approaches unity, it is concluded that distribution is the principal factor in the action of these depressant drugs [28].

For all other kinds of drugs, it is necessary to insert one or more extra terms in the above equation, to provide *multiple* regression analysis [29]. The most frequent additions made by Hansch are the sigma and rho constants from Hammett's Linear Free Energy Equation:

$$\log(K/K_0) = \rho\sigma$$

where K_0 is the ionization constant of benzoic acid (which is the standard) K is the ionization constant of a benzoic acid that bears the substituent under investigation, ρ is a constant pertaining to the nucleus under investigation, and σ (sigma) is the constant pertaining to the substituent. Hammett's Equation is very useful for making a rough prediction of ionization constants in advance of making a selection of compounds for synthesis as candidate drugs. This is useful because classes of drugs with different pharmacological effects have preferred degrees of ionization [30]. For example, local anaesthetics as well as antihistaminics are generally most active if their pK_a values lie between 6 and 8: these constants ensure that workable amounts of both neutral species and cation are present, the former (being more lipophilic) assisting penetration, the latter being the species reacting with the receptor.

A third term often present in a Hansch multiple regression equation is (E), a steric factor which is quite difficult to assess. It is evident that by introducing extra terms of adjustable dimensions, good correlation with the experimental data can always be attained without necessarily contributing to a solution of the problem. In this lies the danger of the multifactorial approach.

The value of multiple regression analysis will be established only if it can save time in indicating, *in advance of synthesis*, drugs of enhanced potency. Essentially, it is a branch of statistics rather than of experimental science. Possibly its greatest value, at present, is to point to neglected areas. For instance, it may show that a desired pharmacological effect increases as σ rises, in a series under study. This information could usefully redirect attention to the synthesis of examples of still higher σ values. Results of high precision are not claimed for the method because of the approximate nature of the many constants used. The custom of taking P from tables instead of determining it experimentally can introduce errors arising from unsuspected hydration of the substance under consideration, cf. [31]. The presence of *ortho*-substituents, so common in drugs, introduces further uncertainties. Finally the exact fit of an agonist on to its receptor, although easily visualized with molecular models, and comprehensible in terms of stereochemistry, is not reducible to a precise term for use in a regression equation. A healthy aspect of these studies is the critical eye with which multiple regression analysts view one another's work [32].

Those with no liking for a mathematical approach often produce good correlations within a given series by concentrating on a single property that is dominant in that series. This may be stereochemistry (especially in nonplanar molecules), chelation, ionization constants, hydrogen bonding (both internal and intermolecular), redox potentials, free-radical formation, surface properties, or covalent-bond changes, with reference to biodegradation or even lethal synthesis.

On the other hand, some workers using mathematics are attempting to correlate molecular orbital calculations (e.g. of electron densities) with biological activity [33].

3.10 Conclusion

Each discovery of a structure-activity relationship has seemed, to its contemporaries, to be a universal explanation of drug action, and it has not always been realized that, had this been true, drugs could evoke only one kind of biological effect. Nowadays we are sufficiently informed to accept different explanations for different biological actions.

References

[1] Bergel, F. and Todd, A. (1937), *J. Chem. Soc.*, pp. 1504–1509.

[2] Blake, J. (1848), *Amer. J. Med. Sci.*, **15** 63–81.

[3] Crum Brown, A. and Fraser, T. (1869), *Trans. Roy. Soc. Edinburgh*, **25**, 151–203 and 693–739.

[4] Ehrlich, P. (1910), *Studies in Immunity*, 2nd. edn., Wiley, New York, 23–47.

[5] Clark, A.J., *Handbuch der experimentellen Pharmakologie* (ed. Heffter, A. and Heubner, W.), Springer, Berlin, **E4**, 228 pp.

[6] Loewi, O and Navratil, E. (1926), *Arch. gesamte Physiol.*, **214**, 678–688.

[7] Dale, H.H., Feldberg, W., and Vogt, M. (1936), *J. Physiol. (London)*, **86**, 353–380.

[8] Ing, H.R. (1936), *Physiol. Rev.*, **16**, 527–544.

[9] de Robertis, E. and Schacht, J. (1974), *Neurochemistry of Cholinergic Receptors*, 150 pp., North Holland, Amsterdam.

[10] Woods, D.D., *Brit. J. exper. Pathol.*, **21**, 74–90.

[11] Brown, G.M. (1962), *J. Biol. Chem.*, **237**, 536–540

[12] Albert, A., Rubbo, S.D., and Goldacre, R.J. (1941), *Nature, (Lond).* **147**, 332–333.

[13] Albert, A., Goldacre, R.J., and Phillips, J.N. (1948), *J. Chem. Soc.*, 2240–2249.

[14] Albert, A., Rubbo, S.D., Goldacre, R.J., Davey, M.E., and Stone, J.D. (1945), *Brit. J. exper. Pathol.*, **26**, 160–192.

[15] Albert, A., Rubbo, S.D., and Burvill, M.I. (1949), *Brit. J. exper. Pathol*, **30**, 159–175

[16] Lerman, L.S. (1964), *J. Cell. Compar. Physiol.*, suppl. 1, **64**, 1–18.

[17] Strugger, S. (1940), *Jena. Z. Naturwiss.*, **73**, 97–134.

[18] Lerman, L.S. (1961), *J. Mol. Biol.*, **3**, 18–30.

[19] Hurwitz, J., Furth, J.J., Malamy, M., and Alexander, M. (1962), *Proc. Nat. Acad. Sci, U.S.*, **48**, 1222–1230.

[20] Albert, A., Rubbo, S.D., Goldacre, R.J., and Balfour, B. (1947), *Brit. J. exper. Pathol.*, **28**, 69–87

[21] Albert, A., Gibson, M.I., and Rubbo, S.D. (1953), *Brit. J. exper. Pathol.*, **34**, 119–130.

[22] Beckett, A.H., Vahora, A.A., and Robinson, A.E. (1958), *J. Pharm. Pharmacol.*, **10**, 160T–170T.

[23] Albert, A., Rees, C.W., and Tomlinson, A.J.H. (1956), *Brit. J. exper. Pathol.*, **37**, 500–511.

[24] Sijpersteijn, A.K., and Janssen, M.J. (1959), *Antonie van Leeuwenhoek*, **25**, 422–438.

[25] Lands, A.M., Arnold, A., McAuliff, J.P., Luduena, F.P., and Brown, T.G. (1967) *Nature (Lond.)* **214**, 597–598.

[26] Choo–Kang, Y.F.J., Simpson, W.T., and Grant, I.W.B. (1969), *Brit. Med. J.*, **2**, 287–289.; Cullum, V.A., Farmer, J.B., Jack, D., and Levy, G.P. (1969), *Brit. J. Pharmacol.*, **35**, 141–151.

[27] Isaki, K., Matsuhashi, M., and Strominger, J., (1966), *Proc. Nat. Acad. Sci., U.S.*, **55**, 656–663.

[28] Hansch, C., Steward, A.R., Anderson, S.M., and Bentley, D. (1968), *J. Med. Chem.*, **11**, 1–11.

[29] Hansch, C. (1968), *J. Med. Chem.*, **11**, 920–924.

[30] Albert, A. (1952), *Pharmacol. Rev.*, **4**, 136–167 (review).

[31] Albert, A., (1967), *Angewandte Chemie*, (Internat. Ed., in English), **6**, 919–928 (review).

[32] Martin, Y.C. (1970), *J. Med. Chem.*, **13**, 145–147; Cammarata, A., Yau, S.J., Collett, J.H., and Martin, A.N. (1970), *Molec. Pharmacol.*, **6**, 61–66.
[33] Kier, L.B. (1971), *Molecular Orbital Theory in Drug Research*, Academic Press, New York, 279 pp.

Suggestions for further reading

Albert, A. (1973), *Selective Toxicity*, 5th. edn., Chapman and Hall, London, 597 pp.
This deals with all the topics discussed in this chapter, but in greater depth.

Porter, C.C. (1970), *Chemical Mechanisms of Drug Action*, Charles C. Thomas, Springfield, U.S.A., 165 pp.
A selection of topics related to drug action, with particular emphasis on biopolymers.

Gale, E.F., Cundliffe, E., Reynolds, P.E., Richmond, M.H., and Waring M.J. (1972). *The Molecular Basis of Antibiotic Action*, Wiley, London, 456 pp.
Five authors, with considerable experience in this field, discuss the title subject.

Goldstein, A., Aronow, L., and Kalman, S.M., (1974), *Principles of Drug Action*, Wiley, New York, 884 pp.
A pharmacological presentation of material discussed in this chapter and the next, with further related topics.

Ariëns, E.J. (editor) (1971–2), *Drug Design*, Academic Press, New York (vols 1–3).
A series of essays on the rational approach to the design and development of biologically active substances.

Some annuals: Advances in Drug Research (Academic Press), *Progress in Medicinal Chemistry* (Butterworths), *Medicinal Chemistry* (Wiley), *Annual Reports in Medicinal Chemistry* (Academic Press) (these last two volumes are sponsored by the Division of Medicinal Chemistry of the American Chemical Society), *Annual Review of Pharmacology* (Annual Reviews, Inc., California), *Pharmacological Reviews* (American Society of Pharmacologists), *Fortschritte der Arzneimittelforschung*, (Birkhäuser, Basel).

4 Favourable differences in distribution: the first principle of selectivity

4.1 Some examples

Many substances which would be toxic for all kinds of cells are nevertheless selective in their action because of favourable differences in distribution. A striking illustration is the use of sulphuric acid to prevent the growth of weeds in cereal crops, as recounted on p. 11. Somewhat similarly, phenothiazine (a well-known oral anthelmintic in sheep) is accumulated by intestinal worms, but not by the cells lining the sheep's gut. Yet, when injected, phenothiazine is toxic for both species [1].

A truly dramatic example from chemotherapy is the accumulation of tetracycline antibiotics by bacteria (Gram-negative and -positive) but not by mammalian cells. As a result, the synthesis of proteins on bacterial ribosomes is drastically repressed and the organisms die. When both the economic and uneconomic cells were fractionated, it was found that the ribosomes of the host were just as sensitive to these antibiotics as those of the parasites. However, the intact cells distribute these drugs so selectively that the tetracyclines constitute one of the most useful series of chemotherapeutic agents [2]. (Other antibiotics act differently).

Examples will now be given of selective partitioning between the tissues of a single organism. The thymine analogue 5-fluorouracil (*4.1*) is used by dermatologists in anti-cancer therapy. Malignant growths, such as basal and squamous cell carcinomas, can be completely eradicated with this drug. So selective is this substance that the patient is encouraged to rub it daily into the affected area, with the bare hand. The cancerous area (only) becomes inflamed, then finally disintegrates and is replaced by normal skin [3].

Another beneficial example of selective accumulation is provided by cyanocobalamin (vitamin B_{12}) which, injected into a muscle, travels to the bone-marrow and is accumulated there after a dilution of 10^{10}-fold in the body fluids. Even a microgram, injected in this way, is enough to bring about the formation of new reticulocytes in the marrow of a pernicious anaemia patient.

Iodine is accumulated by the thyroid gland. When given in radioactive form (^{131}I), 80 per cent of the dose is quickly concentrated by the thyroid; this accumulation has been used selectively to destroy tumours in this gland. Similar studies with ^{32}P have shown that phosphates are specifically taken up by the bone-marrow, and this isotope has proved an effective treatment for polycythemia. Boron derivatives are selectively taken up by the brain [4].

Iodinated derivatives of benzene or pyridine are selectively excreted in the bile (e.g. iopanoic acid), or in the urine (e.g. diodone) (*4.2*), and are much used as X-ray contrast agents in radiography of the gall bladder or kidney, respectively.

Each steroid hormone has a specific protein to transport it to the particular cells whose DNA it has to de-repress. Thus when oestradiol, the natural female sex hormone, was tritiated and injected into the muscles of the arm, it was

5-Fluourouracil
(4.1)

Diodone
(4.2)

found exclusively in the uterus and vagina [5]; this carrier protein has been isolated and purified. Similarly dihydrotestosterone, the natural male hormone, was strongly accumulated by the prostate gland but not by the brain, liver, thymus, or diaphragm. In this case, the carrier protein is so specific that it cannot bind so closely related a substance as testosterone [6].

4.2 Qualitative aspects of distribution

To understand how selective distribution is brought about requires analysis, and this is attempted qualitatively in Fig. 4.1. After being administered, a drug must cross many semi-permeable membranes to reach its receptor. Thus an orally-given antimalarial, chloroquin for example, must penetrate the gastrointestinal membranes, then those of the erythrocyte, and then those of the schizont inside it. Each membrane, because of its restrictive permeability, contributes to the total selectivity. After absorption into the circulation, the concentration of a drug falls through the operation of three factors: storage, inactivation and excretion.

Storage takes place in several ways. Liposoluble drugs (such as the barbiturate thiopentone) are accumulated in deposits of fat; anionic drugs (such as aspirin, sulphonamides, and anticoagulants) by the albumin in the serum; cationic drugs (such as chloroquin) by nucleic acids. If the binding is too tight, these accumulations must be classed with the 'sites of loss', as originally defined by Veldstra [7]. However, when the drug is less strongly bound, and this depends on fine details of its molecular structure, these depots play a useful part in replenishing the plasma level when it begins to fall.

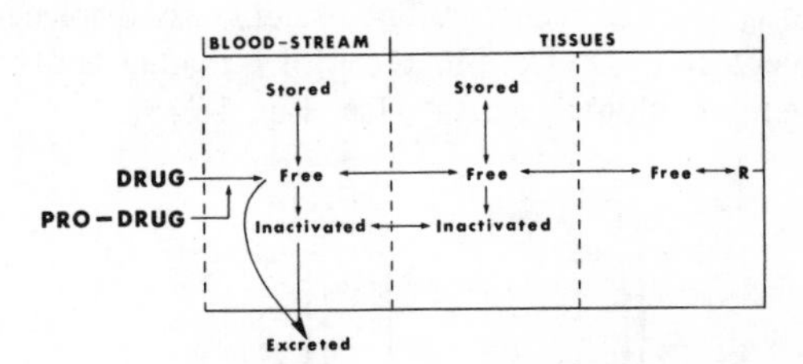

Fig. 4.1 Distribution of a drug in the mammalian body. The broken vertical lines represent semi-permeable membranes, whose presence creates a series of compartments. R stands for the receptor.

Excretion may be through the kidneys, the bile-duct (and hence into the gut), or through the lungs (as with general anaesthetics). Ether and strychnine are examples of drugs which are rapidly excreted without storage or chemical change. For a great many other drugs, excretion is preceded by inactivation, a process that involves the making of breaking of covalent bonds and hence is not freely reversible.

When the drug has penetrated into the tissues through membranes of the blood vessels, it is subject again to loss by storage and inactivation, but in many situations it cannot be excreted until it penetrates back into the blood-stream, a limitation which tends to maintain the concentration. Finally, after surviving all these hazards of storage, excretion, and inactivation, and penetration through other membranes into other compartments, the drug eventually enters the compartment that contains a receptor with which it can bind to provide its characteristic pharmacological response. Union between drug and receptor is covalent in only a few cases e.g. penicillin, mercurials, organic phosphates) and more usually takes place by secondary (easily reversible) bonds. Thus, as the concentration of drug is steadily falling, for reasons given above, steady medication must be maintained to keep an effective concentration of the drug near the receptor. The usually easy displacement of a drug from its receptor is made use of in many biological assays (see Fig. 4.2).

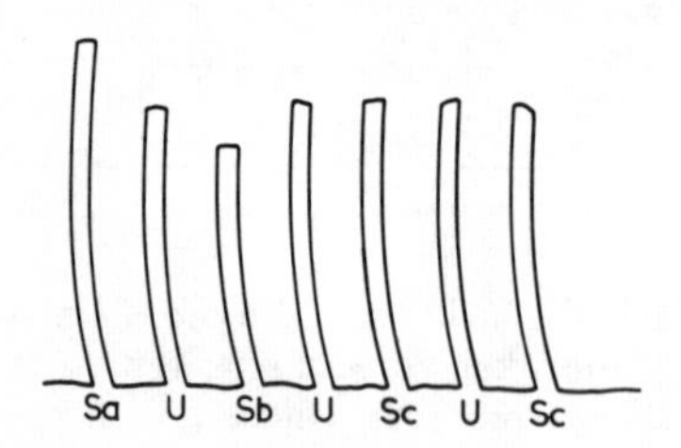

Fig. 4.2 Easy displacement of a drug from receptors by washing. This example is an assay of histamine by immersing guinea-pig gut alternately in unknown (U) and standard solutions (Sa, Sb, and Sc), and washing with saline between each reading until a good match is obtained twice [9].

The concept of apparent volume of distribution (V_D) is often used in these studies, and is defined as:

$$V_D = \frac{\text{Weight of drug in the body}}{\text{Plasma concentration of the free drug}}$$

Thus V_D is the fluid volume in which the drug seems to be dissolved. Values of V_D resembling the volume of a body compartment may suggest that the drug is confined to that compartment. Values of V_D greater than the total body volume indicate that the drug is deposited in a tissue. Some relevant volumes of body compartments are (in litres): circulating plasma of blood (3), erythrocytes (3), extracellular water other than blood (11), intracellular water (24); the total is 41 litres or 58 per cent of body weight [8].

There are also some important recycling mechanisms. Drugs circulating in the bloodstream enter the liver by the hepatic artery and portal vein. The drugs emerge with the bile and pass through the gall-bladder into the duodenum, a slender tube that connects the stomach to the rest of the small intestine. Some drugs, such as phenolphthalein, are absorbed from this intestine by the portal vein, and the recycling continues with a gradual decrement via the faeces. Another cycle follows the route: intestine-lungs-bronchi-throat-intestine with a gradual decrement through expectoration.

The structure of water. The role of water in all distribution phenomena is so important that a few words on its structure seem relevant here. Because water is so familiar to us from our earliest years, we are not inclined to accept the fact that it is the most complex of all known liquids. Its ubiquitous and irreplaceable role in all living processes calls for deeper understanding. Water is not just an inert medium, accidentally present, but has unique physiochemical properties. The curious fact of a maximum density at 4°, and the ability of water to absorb or release calories without much change in temperature, have profoundly influenced the distribution and nature of life on earth. Closer study of its physical properties with modern instrumentation reveals water as a large and complex polymer, existing as such from the melting temperature of ice right up to the transition to steam at 100°.

Liquid water is unique, not only for its powerful facilitation of acid-base equilibria and redox equilibria, but also in its outstanding ability to promote three-dimensional order, for it is the only possible molecule which, from a single atom (O) can give rise to *four* hydrogen bonds (two as donor, and two as acceptor).

When water molecules are in close contact with a paraffin, or the hydrocarbon side-chain of a drug, or an inert gas, the normal quota of four hydrogen bonds obviously cannot be used. This leads to a local gain in density of the water structure. At the other extreme, the structure of water can be opened up with such chaotropic agents as urea or guanidine. These are much used by biochemists to change the conformation of enzymes in contact with water by removing the structure on which they lie extended. Some enzymes are inactivated, and others activated, by this treatment. Just how the properties of water are modified by drugs, and what the biological consequences are, is much discussed at the present time but the main rules have yet to be discovered.

4.3 Quantitative aspects of distribution

The phenomenon of selectivity by distribution is capable of more quantitative treatment. All of the processes that govern distribution can be quantified. To each transition denoted by an arrow in Fig. 4.1, a velocity constant can be attached; storage, excretion, and inactivation rates are being determined for as many drugs as the specialized workers can handle, seemingly a Herculean task.

In such studies, known as pharmacokinetics, the rate of transfer of a drug across a membrane is described by the differential da/dt, where a is the change in amount (or in concentration) effected in a very small interval of time (t). Most pharmacokinetic processes display first-order kinetics. In these, the amount (or concentration) (D) of a drug remaining on one side of the membrane determines the rate of transfer across the membrane, as follows:

$$-da/dt = K_aD$$

where K_aD is the absorption (or distribution, or elimination) constant. For antibacterial sulphonamides, the requisite data may be simply obtained by giving oral doses of one of these drugs to volunteers, and analysing samples of blood and urine at intervals. Such a rate project is outlined in Fig. 4.3 where it was possible to obtain one constant for the storage of the sulphonamide in serum albumin, one for the elimination of the unchanged sulphonamide, one for its acetylation, and one for the elimination of the acetylated product. Each constant is a variable, characteristic of the sulphonamide under investigation. The smallest possible molecular changes in these drugs, such as adding, or subtracting, (or even just moving) a methyl group, significantly alter some or all of these constants, and useful correlations have been established [10].

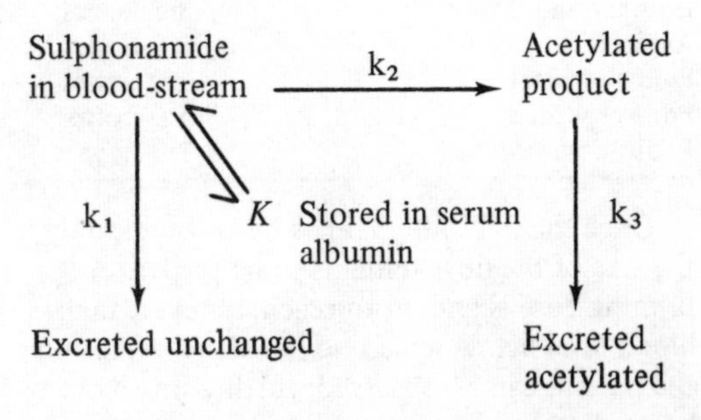

Fig. 4.3 Kinetics of the metabolism and excretion of sulphonamides in the human body [10].

Similarly, for the many other equilibria denoted in Fig. 4.1, rate (and equilibrium) constants are constantly being sought for established, as well as potential, drugs. In this way drug designers are acquiring data to control distribution, and so improve selectivity.

Two other benefits, of great importance, follow from these studies. Firstly the half life ($t_{0.5}$) of the drug in the blood-stream is calculated from the elimination constant by this simple equation:

$$t_{0.5} = 0{\cdot}693/K_a$$

It represents the time taken for half of the drug to disappear from the blood. The wide range for the half lives of various drugs is evident from Table 4.1. Knowledge of this constant (best determined individually for each patient if time and facilities permit) gives the correct dose interval, usually made equal to $t_{0 \cdot 5}$.

Table 4.1 Half-lives of various drugs in the human body [11].

Tubocurarine	13 minutes
Penicillin	28 minutes
Erythromycin	1·6 hours
p-Aminosalicylic acid	1·9 hours
Streptomycin	2·3 hours
Chlorotetracycline	3·5 hours
Imipramine	3·5 hours
Aspirin	5·8 hours
Pentobarbital	42·0 hours
Phenylbutazone	45·0 hours
Bromide anion	7·5 days

Once the dose interval has been found, only the size of the dose remains to be settled. A priming dose is chosen to reach, quickly, that blood level below which no beneficial effect on the patient can be observed; further doses must be smaller, but large enough to prevent the concentration falling below this criterion. If the size of these doses cannot be determined for each patient, by blood analyses performed at the beginning of treatment, they may be taken from tables, of which Table 4.2 is a typical excerpt. The latter procedure is not so fair to the patient, for there are large variations between individuals.

In the early days of treatment with sulphonamides, the wide span of their half-lives was not realized, and examples with long half-lives were often discredited because unwitting overdosage harmed the patient. To-day, modern knowledge of pharmacokinetics prevents the cure being delayed by underdosage, or the patient injured by overdosage. It has also enabled modified drugs with specialized uses to be found with a minimum of useless experimentation.

4.4 The permeability of natural membranes

From the foregoing, it is evident that the distribution of a drug (and hence, in the present

Table 4.2 Recommended dose intervals, and ratios of priming dose (P) to maintenance dose (M), for obtaining steady blood levels in a series of antibacterial sulphonamides [12].

Sulphonamide	*Average half-life (hours)*	*Dose interval (hours)*	*P/M*
Sulphathiazole	3·5	4	1·8
Sulphisoxazole	6·1	6	2·0
Sulphanilamide	8·8	8	2·1
Sulphadiazine	23·5	24	3·0
Sulphamerazine	23·5	24	3·0
Sulphadimethoxine	41·0	24	3·0

context, its selectivity) is highly dependent on its ability to penetrate semipermeable membranes. The classical conception of a biological membrane was that its behaviour was static, something like a dialysis sac. In recent years, latent dynamic properties have been revealed, such as phase-reversal and processes akin to enzymic activity, e.g. permease-based accumulation, and transport in response to metabolism. In what follows, four main types of membrane will be discussed according to their observed permeability behaviour. This will be followed by description of the permeability to drugs of various types of cell and tissue.

Type 1 membranes. These seem to be the most common, or at least they have been found to be the ones that most of the commonly used drugs have to penetrate. Through this type of membrane, those molecules diffuse fastest that have some lipophilic groups, few hydrophilic groups, and no ionic charge. The time for half-equilibrium can vary from under 1 minute to 30 days. Heavy lipophilic substitution can overcome the tendency of an ion not to penetrate, especially if the hydrocarbon substituents crowd around the ionic centre and partly mask it.

The pioneer work in this field was done in Finland by the biologist Collander [13], who also showed that the partition coefficient *P* of a substance in a pair of solvents can be converted to the partition coefficient P' in a second pair of solvents by the following equation:

$$\log P' = a \log P + b$$

where a and b are experimentally determined constants for the particular solvent shift. In recent years, most partition coefficients have been obtained for the solvent pair octanol/water, but values obtained in other solvents can be converted with the help of the above equation [14]. Hansch's π values provide a semi-quantitative idea of the relative lipophilicity of common substituents:

$$\pi = \log P_X - \log P_H$$

where P_X is the partition coefficient of a derivative and P_H that of the parent substance. It may be useful to re-read the latter part of Section 3.8 (p.18) at this point. Table 4.3 lists some π values of substituents inserted in the *m*- or *p*-positions of the benzene ring of phenoxyacetic acid. Values from *o*-substituents are different and unpredictable, and even the listed values differ from those obtained when phenol, or an aliphatic molecule, replaces the acid [15]. π-Values are roughly additive.

Few molecules penetrate a Type 1 membrane if they have more than two hydrophilic (e.g. —OH) groups. The involvement of a type 1 membrane is highly likely if various substances, of roughly the same molecular weight and diameter, are found to penetrate a membrane at a rate proportional to their partition coefficients.

For a discussion on membrane permeability and diffusion, see Wilbrandt [16]. The cardiac glycosides afford clinically interesting examples of the effect of partition coefficients on permeability. Digitoxin, the most readily accumulated glycoside in clinical use, is the most lipophilic. It is excreted only slowly into the bile and is largely resorbed from this fluid Related glycosides that are more hydrophilic because of the presence of extra sugar groups, or extra hydroxyl groups in the steroid nucleus, are more readily lost in the bile (e.g. digoxin and lanatoside C).

Table 4.3 Relative lipophilicity of substituents, inserted into the benzene ring of phenoxyacetic acid (Hansch's π values) [15].
(The higher the value, the more lipophilic)

Substituent	*π-value*
H	0
Cl	0·73
I	1·20
Me	0·52
Bu	1·90
C_6H_5	1·89
NO_2	0·17
SMe	0·62
OMe	0·08
CO_2H	−0·15
COMe	−0·33
CN	−0·31
OH	−0·55
SO_2Me	−1·26

Type 2 membranes. These incorporate a carrier which provides facilitated diffusion. The transported molecule combines reversibly with the carrier, which oscillates between the inner and outer surfaces of the membrane to release or pick up this molecule. The tests for this type of membrane are (a) the barrier can become saturated, whereupon transport may cease, even though the gradient remains favourable, and (b) no metabolic energy is consumed by the act of transport (e.g. no increase in respiration can be detected).

Facilitated diffusion of this kind is provided for choline by erythrocytes and nerve cells. Choline, being a quaternary ammonium salt, is permanently ionized and hence cannot enter by simple diffusion, such as a Type 1 membrane requires. Of analogous foreign substances, tetramethylammonium ions can enter cells on the choline carrier, whereas higher tetra-alkylammonium salts cannot, but they do block the physiological uptake of choline [17].

The transport of glucose into human erythrocytes is the most studied example of facilitated diffusion; it is thought that the carrier esterifies the 2-hydroxy-group. At least seven carriers control entry through the membranes of mitochondria. One carrier facilitates entry of succinate, malate, malonate, and meso-tartrate anions, but not of tartrate, maleate, or fumarate. Another carrier mediates the entry of citrate, *cis*-aconitate, and tartrate. A third carrier transports adenosine nucleotides. Also phosphate anions can enter mitochondria whereas other inorganic anions cannot [18].

By designing drugs to contain groups essential for the facilitated uptake of natural nutrients, increased selectivity could be expected. Whole classes of medicaments, less lipophilic than currently used drugs usually are, might be anticipated.

Type 3 membranes. These, the most complex of all, operate an energy-consuming process, and hence can concentrate particular substances against a gradient if necessary. The penetrating molecule is thought to combine with a carrier, as in Type 2, but this carrier is subjected to chemical modification in the membrane. Inorganic cations penetrate into mammalian cells, and iodide anion into the thyroid gland, in this fashion; also the absorption and excretion of a wide range of ionized and non-ionized substances by the kidney tubules and, to a less extent, by the gastro-intestinal membrane is effected in this way. For each 18 sodium ions transported through the intestinal wall, one extra molecule of oxygen is consumed. Many laxatives (e.g. cascara, phenolphthalein, podophyllin) inhibit absorption of sodium by the membrane of the intestinal lumen (found from experiments in living rabbits) and hence cause accumulation of sodium salts in the colon which leads, through retention of water, to evacuation.

The type 3 membrane, which seems to occur in patches on a normal Type 1, can be recognized by (a) the ease of saturation of the carrier (as in type 2), and (b) the increased metabolic activity during transport.

When new drugs are being designed, the requisite permeability has usually been obtained by stepwise increases in lipophilicity, a strategy suitable for Type 1 membranes. A different approach is to use Type 2 and 3 membranes by making the drug resemble, in the relevant part, a natural substance for which a specific transport mechanism exists in particular cells. (To this end, it is interesting to read how much change is permissible in the molecule of glucose before the Type 2 carrier fails to accept it [19]). That this approach can be highly successful is shown by its ability to increase the penetration of nitrogen-mustard anticancer drugs [20]. Various aminoacid groups were attached, and the phenylalanine derivative penetrated particularly well; under the name melphelan it is giving excellent results in the clinic. Other workers report promising results by attaching natural sugars.

Type 4 membranes. These are distinguished from type 1 by the presence of pores, which in the glomerular tuft of the kidney are as large as 3 nm in diameter: through these even as large a molecule as inulin (mol. wt. 5000) passes easily. Small anions, such as chloride, seem to pass into cells through similar channels lined with positively charged groups; these channels exclude cations by coulombic repulsion. No serious attempts have yet been made to use type 4 membranes to improve the selectivity of drugs.

Pinocytosis. This is a process in which molecules, especially proteins, too large to diffuse through a cell in the normal way, are caught up in an invagination (of the cytoplasmic membrane) which is then pinched off to form vesicles. By this means, large molecules formerly outside the cell can appear within it, and vice versa. The proximal tubules of the kidney resorb protein in some such way while engaged in discriminating between many kinds of smaller molecules.

Phagocytosis. This is a similar process which effects the passage of pre-formed vesicles. In this way, nerve-endings secrete vesicles which contain the neurotransmitter acetylcholine.

Ionization. For the better understanding of what follows, a few words of revision on ionization will be inserted. It will be recalled that the pK_a of an acid or base is the negative-logarithm of the ionization constant (i.e. the logarithm of the reciprocal of the ionization constant). The stronger an acid is, the lower the pK_a; the stronger a base is, the higher its pK_a. The pK_a, it is useful to remember, is numerically equal to the pH at which a substance is 50% ionized. When the pH is one unit below the pK_a, an acid is 9% ionized (a base, 91%). When the pH is one unit above the pK_a, an acid is 91% ionized (a base, 9%). Each acid or base has its own intrinsic pK_a, but the extracellular pH can be altered to suit the experimenter. In a sense, the pK_a can also be altered by making a change, often quite a small change, in the structure of the molecule guided by knowledge of Hammett sigma values (see p.19). Some pK_a values of common substances, useful for making comparisons, are: hydrochloric acid (<0), acetic acid (5), phenol (10), glucose (13); aniline and pyridine (5), most alkaloids (8), ammonia (9), sodium hydroxide (>14).

The degree of ionization (in aqueous solution) of any base can be calculated from the following equation:

$$\text{per cent ionized} = \frac{100}{1 + \text{antilog}\,(\text{pH} - pK_a)}$$

It is seen from this equation that the degree of ionization varies with pH, but follows a sigmoid, rather than a straight-line, relationship. Thus a small change in pH can make a large change in the percentage ionized, particularly if the pH of the solution is close to the pK_a of the substance under investigation. See Ref. [22] for more information on ionization, for tables of pK_a values, and for a table giving percentage ionized when pH and pK_a are known.

The permeability of mammalian tissues. The absorption and distribution of foreign substances follows a much simpler pattern than those of natural cell substrates and active constituents. The following structures offer a Type 1 lipid barrier to the passage of many foreign molecules: the gastro-intestinal epithelium, the renal tubule epithelium, the blood-brain barrier, and the blood-cerebrospinal fluid barrier [21].

From the stomach, drugs are well absorbed only if not ionized. For example, when the pH of the stomach contents was raised, basic drugs were found to be better absorbed because they were less ionized; but this pH change decreased the absorption of acidic drugs because less of the non-ionized form remained.

The value of lipophilic properties in assisting absorption can be demonstrated with three barbiturate hypnotics of similar pK_a but different lipophilicity (see Table 4.4). where A is the per cent absorption from a rat's stomach when fed a solution (at pH 1) orally, and P_c is the partition coefficient (the higher the value of P_c, the more lipophilic the substance).

Table 4.4 Correlation of gastric absorption with lipophilicity [23].

Barbiturate	pK_a	A	P_c
Barbital	7·8	4	<0·001
Quinalbarbital (secobarbital)	7·9	30	0·10
Thiapental	7·6	46	3·30

The pattern of absorption from the human stomach follows the principle set out above. Weakly acidic drugs, such as salicylic acid, aspirin, and the more lipophilic barbiturates, were readily absorbed because they were not ionized at the pH of the stomach (pH 1); whereas basic substances, such as quinine, ephedrine, and amidopyrine, were not absorbed because they were totally ionized at this pH.

The small intestine's epithelial lining was similarly found to permit the penetration of non-ionized drugs and impede the passage of the corresponding ions. The average pH (6·6) of the small intestine, so much higher than that of the stomach, permits the passage of aromatic amines (pK_a usually 5), but not of the stronger aliphatic amines (pK_a about 11). Raising the pH increases the absorption of bases and decreases that of acids, just as in the stomach. Moreover, in both organs, substances whose degree of ionization is not changed by such alterations in pH show no change in the rate of absorption. As in the stomach, increasing lipophilicity helps absorption.

Absorption of natural substances from the small intestine, e.g. aminoacids, glucose, and uracil, uses specific activated transport systems which can work against a concentration gradient. Foreign substances that closely resemble such natural substrates can be transferred from small intestine to plasma in this way.

The absorption of drugs from the colon was found to be similar to that from the small intestine. The distribution of drugs between blood-plasma and the tissues was essentially the same as between the gastrointestinal tract and the blood plasma. However, only the fraction *not* bound to plasma protein is free to diffuse in this way (cf. p. 23). A weakly bound drug can be displaced by one for which serum albumin has a higher affinity. Neutral organic substances diffuse across the skin proportionally to their lipophilicity.

The capillaries of the blood-system and those of the kidney glomerulus have porous Type 4 membranes which allow the passage of ions and small proteins. The human kidney produces 185 litres of glomerular filtrate a day, but the renal tubules absorb all but 1·5 litres of water and also many dissolved substances of great value in the body's economy. These tubules have normal Type 1 membrane with specialized patches for activated transport; the excretion of foreign organic anions is most efficiently brought about in this way.

Of two similar drugs, the one that is more strongly bound to blood-albumin is excreted by the glomerulus with more difficulty. This means that porosity is of little avail against the secondary valencies that tie a drug to a blood-protein. On the other hand, protein-binding is of little avail against excretion if the constitution of the drug makes it eligible for the active transport mechanisms of the renal tubule. These principles are relevant when redesigning a drug with a view to reducing the effective dose.

Molecules much larger than those of the commoner antibiotics and synthetic drugs, even inulin (mol. wt. 5000), pass readily from blood into bile. Substances of mol. wt. under 400 are poorly excreted into bile, but an anionic group helps. This important pathway for the absorption and excretion of drugs is insufficiently understood. The liver has large pores (without parallel in other animal cells) in the outer membrane of hepatic parenchymal cells.

Drugs penetrate from the blood into the cerebrospinal fluid at rates parallel to their lipid/water partition coefficients. The only ionized substances that penetrate easily are the phenylarsonic acids used in treating trypanosomiasis: these, it is thought, are taken up by the permease for phosphate transport. Regardless of their rate of entry into the CSF, drugs often disappear quite rapidly from this fluid, a phenomenon needing more investigation.

The brain is protected by a membrane, often called the blood-brain barrier, which lines the capillaries of the cerebral vessels. This type 1 membrane can be penetrated only by substances with a very high lipid/water partition coefficient, but the brain can out-compete all other tissues for these highly lipophilic drugs. After passing this barrier, the drug still has to penetrate the membranes of the brain. Whereas sulphonamides and the more lipophilic of the barbiturates readily penetrate the blood-brain barrier, acetanilide and barbital penetrate only slowly, and penicillin and the salicylates not at all. However, when the membranes are inflamed, a wider range of drugs can pass through.

Transfer of drugs has been found to occur between different regions of the brain. Highly lipophilic substances were found in the very vascular grey matter soon after oral administration; but a few hours later they were seen to be localised in the white regions which are richer in lipids.

Barriers to permeability *within* cells consist mainly of the membranes of mitochondria and the endoplasmic reticulum, both of which appear to be type 1 with patches of types 2 and 3; and also the nuclear membrane which appears to be the highly porous type 4.

Bacteria, whether intact or freed from cell walls, have, on the exterior of the cytoplasm, a mainly type 1 membrane which is their sole permeability barrier. This has been studied in, for example, *Staphylococcus aureus, Micrococcus lysodeikticus* and *Sarcina lutea* [23a]. Whereas lysine can penetrate this membrane by simple diffusion the more hydrophilic aminoacids, such as histidine and glutamic acid, penetrate only by active transport [23b], just as glucose does.

The chitin membranes of arthropods are type 4.

The membranes of plant cells are remarkably like those of animals. The cytoplasmic membrane of roots seems to have a special structure with special selectivity. However, in general, the plasma membrane of plants seems to offer not so great a barrier as those of the mitochondria, or of the endoplasmic reticulum, or of the vacuole (a structure absent in animals).

Maleic hydrazide causes breaks in the chromosomes of plant cells, but not in those of mammals. Because both kinds of chromosome have the same chemical composition, a difference in permeability is indicated.

In conclusion it can be pointed out that when lipid/water partition coefficients govern uptake or excretion, they seem to have different minimal values in different tissues. Specific binding substances within cells, although little explored, seem also to play an important part. For further reading on the movement of molecules across membranes, see Ref. [24].

4.5 Metabolic change leading to activation

A glance at Fig. 4.1 will remind us that some drugs are not administered as the form in which they react with their receptors, but as pro-drugs which have to be converted to the true drug by enzymes present in the body. This masking has arisen intentionally in some cases, accidentally in others. It is not uncommon, but most drugs appear to act in the form in which they were given; but it is a process with great potentiality for increasing selectivity.

The first pro-drug to be designed as such was hexamine (urotropin, methenamine) which was introduced in 1899 as a source of formaldehyde, liberated from it by the acidity of urine. The anthracene glucoside purgatives had been used for centuries (in crude forms, as cascara, senna, rhubarb, etc.) before they were recognized as pro-drugs: their aglycones (e.g. emodin) are the true active forms. Sodium citrate which, after oxidation to sodium bicarbonate, basifies the urine is another early example of a pro-drug.

Of drugs introduced in the latter half of the 19th century, it is now known that chloral hydrate is reduced in the body to trichloroethanol which is no less hypnotic. also acetanilide and phenacetin act as analgesics only after metabolism to *p*-acetamidophenol which is now prescribed instead of these two pro-drugs.

Early in this century, Ehrlich found that phenylarsonic acids, e.g. (*4.3*) were inactive against trypanosomes until converted in the body to the corresponding arsenoxides. In treating trypanosomiasis, it was found useful to let phenylarsonic acids diffuse into the site of infection in the central nervous system, and there become reduced to the true drug, because arsenoxides, e.g. (*4.4*) do not penetrate the blood-brain barrier. In this example, the masking of the arsenoxide (by administering it as the corresponding arsonic acid) was advantageous. Such was not the case for the arsenobenzenes, e.g. (*4.5*), which act only after oxidation to arsenoxides (*4.4*). The arsenobenzenes, e.g. Ehrlich's historic antisyphilitic drug salvarsan, have to be given in much larger doses than the arsenoxides which later replaced them [25] because their oxidation was highly inefficient; moreover it led to arsenical by-products toxic to the patient. Here, in closely related series, is an example where masking increases selectivity and one where it decreases it.

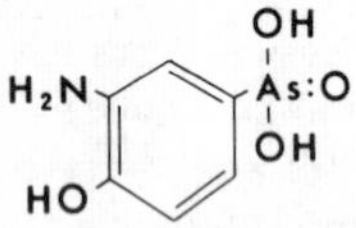

An arsonic acid
(*4.3*)

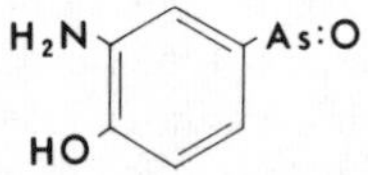

An arsenoxide
(oxophenarsine)
(*4.4*)

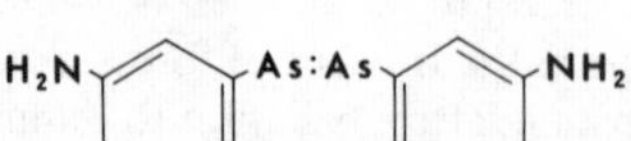

An arsenobenzene
(Salvarsan)
(*4.5*)

'Prontosil' (*4.6*), the first of the antibacterial sulphonamides, was thought to be the true drug

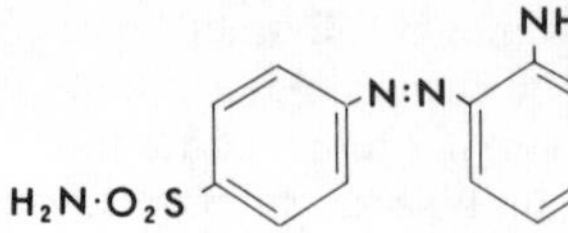

'Prontosil'
(*4.6*)

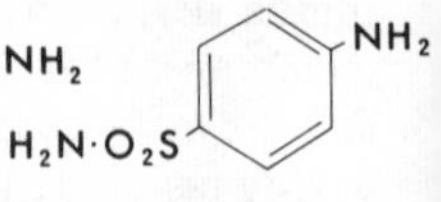

Sulphanilamide
(*4.7*)

when it was introduced into medicine in 1935. But workers in the Institut Pasteur (Paris) were able to show, in the same year, that this substance had no intrinsic antibacterial properties, and that the true drug was sulphanilamide (*4.7*), which was formed by reductive fission in the patient's bowel [26]. Accordingly sulphanilamide replaced 'Prontosil' in the clinic because it acted more promptly and directly.

Whenever no correlation can be found between plasma concentration and therapeutic effect, the 'drug' is probably a pro-drug. A clue of this kind led to the discovery that the antimalarial proguanil ('Paludrine') (*4.8*) acted only after cyclization in the body to the dihydrotriazine (cycloguanil) (*4.9*).

N-Demethylation of foreign substances takes place very readily in the liver, and many pro-drugs (provided with *N*-methyl groups to assist penetration) are converted there to the active form. This device, which increases the speed of action, is built into the following hypnotics and antiepileptics: mephobarbital, hexobarbitone, methoin, troxidone, and paramethadione.

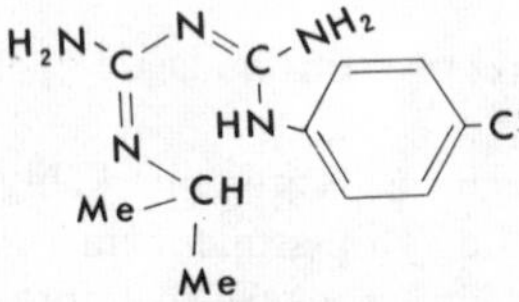

Proguanil
(*4.8*)

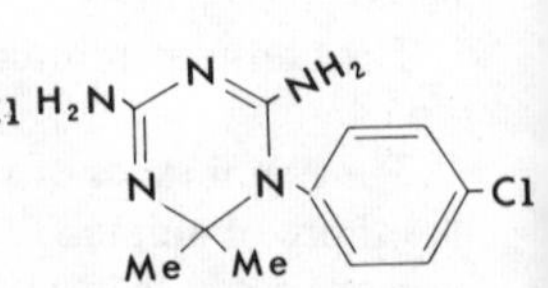

Cycloguanil
(*4.9*)

When a drug is metabolized or excreted too fast for effective medication, it is often converted into a masked form which will form a depot. Examples include penicillin (made into the insoluble procaine salt), and testosterone (as the phenylpropionic ester, dissolved in oil). Succinylsulphathiazole, a valuable remedy for bacterial dysentery, differs from sulphathiazole in that it is poorly absorbed from the small intestine. Accordingly, this pro-drug passes on to the colon where it is hydrolysed by benign bacteria to sulphathiazole, which suppresses the pathogens. Cycloguanil (*4.9*), re-introduced as an insoluble salt ('Camolar'), is being used as a depot antimalarial: a single intramuscular dose can protect a man against malaria for five months.

An ingenious use of masking, to overcome a transport problem, was to make 6-azauridine into its more liposoluble triacetyl derivative for oral medication in cancer. Whereas 6-azauridine does not leave the intestine, the triacetyl derivative is well absorbed; rapid deacetylation liberates the true drug [27]. Somewhat similarly, 5-hydroxytryptophan (which, like most amino-acids passes the blood-brain barrier by a Type 3 mechanism) is decarboxylated in the brain to 5-hydroxytryptamine (serotonin) a substance which cannot enter directly. In the treatment of parkinsonism, dopa (dihydroxyphenylalanine) is administered in order to accumulate the true drug (dopamine) in the brain, after the local decarboxylase has acted on the pro-drug which penetrates so much more readily.

A contemporary trend is to introduce substituents into a drug which can electronically deactivate its functional group. The substituent must be one that will be removed metabolically. To this end, attempts are being made to deactivate the chlorine atoms in the nitrogen-mustard class of anticancer drugs, using deactivating groups known to be removed by enzymes present abundantly in the cancer cells.

Agricultural agents. DDT (chlorophenothane, or dicophane) has been made selective against those cellulose-digesting insects that bite plants, by forming it into granules coated with cellulose. One highly selective insecticide, *N*-methyl-*N*-(l-naphthyl) fluoroacetamide ('Nissol') (*4.10*), is lethal to mites because they liberate fluoroacetic acid from it, but has little toxicity to mammals which do not degrade it in this way.

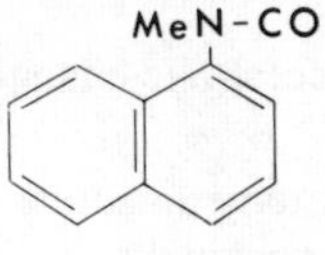

MeN-CO·CH$_2$F

Masked form of fluoroacetic acid (*4.10*)

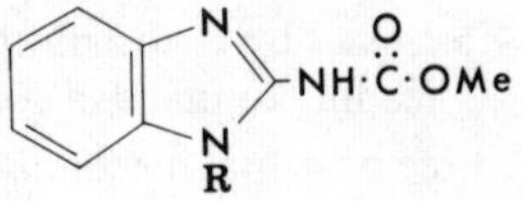

N, NH·C(O)·OMe, N, R

(a) R = C(O)·NHC$_4$H$_9$
(b) R = H
(*4.11*)

The systemic fungicide benomyl (*4.11a*), introduced in 1966, was soon found to be immensely more active and selective than anything else. This substance (methyl l-butylcarbamoyl-2-benzimidazolecarbamate) easily loses the butylcarbamyol group (which may assist penetration into the plant) to give methyl 2-benzimidazolecarbamate (BCM) (*4.11b*), which has the same degree and range of activity [28].

A most ingenious way has been found to increase the selectivity of herbicides in the phenoxyacetic acid series. Many weeds can degrade the side-chain of harmless 2, 4-dichlorophenoxy derivatives of the aliphatic acids (this substituent is inserted at the far end of the aliphatic chain) to form 2, 4-dichlorophenoxyacetic acid, which kills them. Fortunately many economic crops lack the β-aliphatic oxidase necessary for the lethal degradation. Using this principle, excellent selectivity is obtained with dichlorophenoxybutyric acid which can, for example, eliminate flax from a clover crop 129].

Conclusion. Although, from time to time, a substance thought to be an agent turns out to be only a pro-agent, most agents have been found to act on their receptors in the same chemical form as that in which they are administered. A detailed knowledge of permeability and the naturally-occurring degradative enzymes can assist a skilful designer in finding useful pro-agents. However, he should bear in mind that he is adding more complications to the already long list of distribution problems that lie between administration of the agent and its arrival at the receptor.

4.6 Losses by re-distributions. Metabolic change leading to inactivation.

Three principal mechanisms whereby an active substance can be lost before it reaches the effective receptor are: storage, elimination, and chemical inactivation (Fig. 4.1). Veldstra referring to the loci where these processes occur as 'sites of loss', suggested that the well-known synergistic action of biologically inert substances is actually a blockade of such sites, which allows a higher concentration of the drug to reach the receptor [7]. Of the three important *storage sites* (see Section 4.2), *lipids* (represented by body-fat) store drugs of high liposolubility, e.g. thiobarbiturates, dibenamine, and dibenzyline. This appears to be a simple lipid/water partition effect. Plant-growth accelerators, such as α-naphthylacetic acid, are taken up by the fatty reserves of pea shoots; these sites of loss can be blocked with a biologically inactive, but more liposoluble, analogue such as decahydronaphthylacetic acid [7].

Ribonucleic acid combines with the cations of highly basic substances. Thus the principal storage of the highly basic antimalarials, such as mepacrine is in the nuclei of capillaries where it does no harm and is available for replenishing the blood-level. Chondroitin and other anionic biopolymers also store cations.

Several proteins in the mammalian bloodstream are capable of binding drugs, but *albumin* is by far the most effective of these. Neither fibrinogen nor the γ-globulins combine with drugs; and α- and β-globulins are usually enzymes whose affinity is almost confined to their substrates (although β_1-globulin combines with iron, zinc, and copper). Lipoproteins are related to the globulins and combine with steroid hormones (lipid-lipid attraction). But these substances are natural metabolites, and the trypanoicide suramin seems to be the only drug that is bound by a globulin.

Serum albumin, on the other hand, is a storage site for many drugs, most of which are weak acids. Table 4.5 shows how this affinity varies not only from species to species, but also among two pairs of chemically related substances. Man, who binds drugs by serum albumin more strongly than other mammals do, usually metabolizes drugs less readily than other mammals.

Table 4.5 Binding of Drugs by Serum Albumin [30]

	Percentage unbound			
Animal species	*Benzyl-penicillin*	*Cloxacillin*	*Sulpha-diazine*	*Sulph-isoxazole*
Man	49	7	67	16
Horse	59	30	–	–
Rabbit	65	22	45	18
Rat	–	–	55	16
Mouse	–	–	93	69

Human serum albumin, of which blood contains about 4 per cent, has a mol wt. of 69 000, and possesses 109 cationic and 120 anionic groups. Although, at pH 7·3, it contains a net negative charge, one of the cationic groups must be particularly accessible because this albumin binds mainly anions, and in a 1 : 1 ratio. This binding follows the law of mass action. Unless more than 95 per cent of the drug is bound by albumin, renal clearance of the drug is not slowed, and the serum protein must be regarded as a helpful depot and not as a site of loss. The dissociation constant of the drug-albumin complex varies greatly, e.g. 900 for sulphadiazine (which is poorly bound) to 11 for sulphadimethoxine ('Madribon') which seems to be too well bound to be a useful drug.

The clinical effect of a drug is increased, sometimes dangerously, if a second drug displaces it from the albumin. Thus aspirin chases anticoagulants out of their store in serum albumin and often precipitates a crisis of bleeding.

The drugs most readily bound by serum albumin are aliphatic acids; the longer their hydrocarbon chain, the more strongly they are bound (by van der Waals forces). Almost as tightly bound are aromatic acids, sulphonamide drugs, and the barbiturates, particularly if they carry lipophilic substituents (as discussed in Section 3.8). A few neutral substances combine with serum albumin: naphthoquinones, coumarins, indanediones, lactones including the cardiac glycosides, and porphyrins. Very few cationic substances become bound to serum albumin. Among the many simple drugs and metabolites *not* bound by albumin are ether, glucose, and urea.

It has repeatedly been demonstrated that the level of free drug in tissue fluids is the same as that in plasma. This relationship is convenient to know, because the plasma is more assessible for analysis.

Metabolic alteration of drugs involves the making or breaking of covalent bonds and hence is less reversible than the process of storage. Although many hydrophilic drugs are excreted unchanged by the mammalian body, others are 'conjugated', i.e. they are joined to small metabolites to assist excretion. For example, organic acids that are too weak to be ionized at the low pH of urine, and which would therefore be difficult to eliminate, are conjugated with glycine, and are then readily excreted by the kidney. A few amines are acetylated (by acetyl coenzyme A, in liver, gut, spleen, and lungs), but most of them follow the same course as the phenols by being conjugated with glucuronic acid (in liver, kidney, or gut) or sulphuric acid (in the cell sap of kidney, liver, or gut). Glucuronides can also be formed by alcohols, amides, and carboxylic acids but, as the process takes place in the endoplasmic reticulum, they must have a certain liposolubility to be candidates.

Drugs that are yet more lipophilic are resorbed by the renal tubules. If they were not submitted to metabolic degradation, a single dose would circulate in the body for many weeks, out of all therapeutic control. However they accumulate in a membranous organelle in the liver cells known as the *endoplasmic reticulum*, which will be abbreviated to *e.r.* in what follows. This organelle contains many kinds of scavengeing enzymes which alter the substrates chemically to make them more hydrophilic, and readily excreted. For example, toluene is oxidized by the *e.r.* to benzyl alcohol, the oxidation of which is continued in the cytoplasm to give benzoic acid, which is then conjugated with glycine in the mitochondria, and the resultant benzoylglycine (hippuric acid) is rapidly eliminated in the urine.

Although this example shows that *e.r.* is not the sole site of metabolic degradation, this organelle is by far the most versatile. It can be separated from liver by differential centrifugation and freed from the neighbouring ribosomes which are sites of protein synthesis. During the course of such purification, the *e.r.* is broken up into spherules ('microsomes') without loss of enzyme activity. These enzymes, which are numerous, are mainly oxidative, but a few of them perform reductions, hydrolyses, and at least one synthesis [31]. The following typical oxidative processes are performed by these enzymes (at least one enzyme for each process).

(i) Aliphatic *C*-hydroxylation ($R \cdot CH_3 \rightarrow R \cdot CH_2OH$), for which the side-chains of barbiturates are common substrates.

(ii) Aromatic *C*-hydroxylation, e.g. the conversion of acetanilide to *p*-hydroxyacetanilide.

(iii) *N*-Oxidation ($R_3N \rightarrow R_3NO$), for which both aliphatic and aromatic tertiary amines are good substrates.

(iv) *S*-Oxidation ($R_2S \rightarrow R_2SO$), as in the oxidation of chloropromazine.

(v) *O*- and *S*-Dealkylations (e.g. $ROC_2H_5 \rightarrow ROH + CH_3CHO$), for which phenacetin and codeine are well-known substrates.

(vi) *N*-Dealkylation ($RNH \cdot CH_3 \rightarrow R \cdot NH_2 + H \cdot CHO$), as in the conversion of methylaniline to aniline.

(vii) Deamination ($R \cdot CH(NH_2) \cdot CH_3 \rightarrow R \cdot CO \cdot CH_3 + NH_3$), as in the metabolism of the side-chain of amphetamine.

Brodie [32] has convincingly argued that the enzymes of the *e.r.* exist for the degradation of toxic substances normally occurring in food or produced by bacterial decomposition in the gut. In addition, they have a normal role to play in steroid metabolism, particularly in the hydroxylative destruction of such hormones as oestradiol, testosterone, progesterone, and the corticoids. The *e.r.* enzymes can act on a wide range of structures and hence are capable of attacking drugs not previously encountered.

Apart from *N*-oxidation, most of the above reactions require a special cytochrome coenzyme known as P_{450} (because it absorbs visible light intensely at 450 nm), also the coenzyme NADP, and a flavoprotein enzyme (cytochrome c reductase) which utilises the oxygen of air. These *e.r.* enzymes, by their requirement for NADP, stand apart from the many NAD-requiring enzymes that the body's intermediary metabolism uses in its stepwise conversions of nutriment into energy. Conversely, the *e.r.* enzymes cannot attack either the raw materials or the products of intermediary metabolism, because such substances are too hydrophilic to penetrate into the *e.r.* These data afford us a rare glimpse of how selectivity is achieved in Nature.

Metabolic alteration of foreign substances has often been called 'detoxification', but many examples are known where the product of an *e.r.* enzyme is more toxic than the substrate. Thus dimethylnitrosamine is converted to a substance which methylates the guanine of RNA, a reaction which leads to acute liver necrosis. Recently much interest has been shown in another toxifying reaction of the *e.r.* enzymes: many foreign substances with a benzene or naphthalene ring are converted (although only a small proportion is affected) to a poisonous epoxide, e.g. (*4.12*) from benzene, which can cause centrolobular necrosis in liver [33]. This, or a similar, toxic oxidation by the *e.r.* is thought to underlie the carcinogenic action of hydrocarbons. The toxic action of methanol, including eventual blindness, is caused by its transformation by the body into formaldehyde.

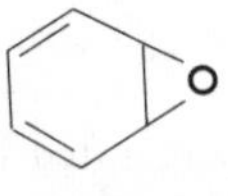

O

Toxic epoxide
(*4.12*)

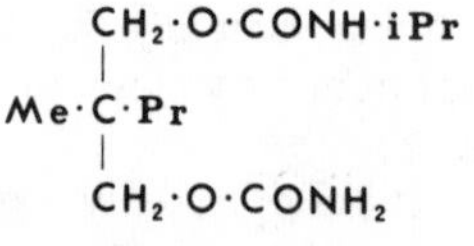

$CH_2 \cdot O \cdot CONH \cdot iPr$
$Me \cdot C \cdot Pr$
$CH_2 \cdot O \cdot CONH_2$

Carisoprodol
(*4.13*)

For most experiments on the enzymes of the *e.r.*, the source has been rat liver. However, it has been shown that human *e.r.* enzymes are qualitatively similar to those of the rat, but act at different rates, some faster and some slower. It has been suggested that the large differences in effective dosage that exist between man and laboratory animals depend more on such species differences in the rate of destruction than on any species differences in the sensitivity of target organs. Hence a given pharmacological effect should appear at a similar blood-level in all mammals, even though the doses required to produce this level are known to vary greatly from one species to another. Table 4.6 supports this argument; it shows some quantitative data on carisoprodol (*4.13*) a muscle relaxant. However, too little information is available to say how widely the above hypothesis would hold.

Table 2.4 Species differences and similarities in the action of a drug, carisoprodol, given intraperitoneally (0·2 g/kg) [34].

Species	*Duration of action (loss of righting reflex) hours*	*Plasma level on recovery μg/ml*
Cat	10	125
Rabbit	5	100
Rat	1·5	125
Mouse	0·2	130

Exceptions to the above hypothesis obviously exist in those cases, possibly rare, when the major pathway of metabolism is not the same in two species.

This possibility of species variations is recognized also in the narrower field of conjugation. Phenylacetic acid, the classic example of divergent paths, is conjugated with glutamine in man and other primates, with glycine and with glucuronic acids in most other mammals and with ornithine in the hen. Again, amphetamine is metabolized in the rat by *para*-hydroxylation, but by deamination in man, monkey, and guinea-pig [35]. Interestingly, rabbits can eat belladonna leaves with impunity because they have an esterase in the blood-serum sufficiently non-specific to hydrolyse the tropine alkaloids, but other mammals lack such an enzyme.

For further reading on metabolic alteration of drugs, see Ref. 36, and the journal *Xenobiotica* which began in 1971.

4.7 Synergism and antagonism

Synergism. Much light has been thrown on one important aspect of the synergism of drugs by the use of the substance SKF 525-A (*4.14*). This substance, the diphenylpropylacetic ester of diethylaminoethanol, can synergize the action of a wide variety of drugs by preventing their metabolism in the *e.r.* It seems to exert this effect, not by making the membrane of the *e.r.* impermeable to lipophilic drugs, but by non-competitive inhibition of all hydroxylation reactions and by competitive inhibition of hydrolytic reactions [31]. These actions of SKF 525-A are examples of a very common type of synergism, namely, blocking sites of loss [7]. It is not used medicinally.

Metabolic inactivation, whether taking place in the *e.r.* or at other sites, is often accidentally inhibited by other drugs. Thus many patients have suffered as a result of the simultaneous administration of an ihhibitor of monoamine oxidase (an enzyme present in mitochrondia) and an amine drug which is not toxic on its own. Thus the inhibitor phenelzine (phenylethylhydrazine) has caused deaths after usually safe doses of amphetamine, pethidine, nortriptyline, or amitriptyline, or after the patient has consumed amine-rich food such as cheese, red-wine, meat-extract, yeast-extract, or broad beans. These are examples of unfortunate synergism, but many favourable examples are known, examples of which will now be given.

Loss by elimination can sometimes be blocked by an analogue of similar charge type. Thus the penicillins belong to the class of moderately liposoluble acids which get facilitated transport through the proximal tubules of the kidney. This elimination can be largely blocked by physiologically inert substances of similar physical properties, such as probenecid ['Benemid', *p*-(dipropylsulphamoyl) benzoic acid]. This substance is used clinically to increase the action of penicillins when a highly resistant strain of bacterium has to be eliminated.

Loss by enzymic destruction can be overcome through use of a synergist. Thus pyrethrins in fly sprays are commonly formulated with a methylenedioxybenzene synergist, often derived from piperic acid. These substances inhibit an *e.r.* enzyme which oxidizes pyrethrins, carbamates, and organic phosphates in the insect body [37]. One of the most used synergists of this class is piperonyl butoxide (*4.15*), but even quite simple methylenedioxy-derivatives of benzene show the effect.

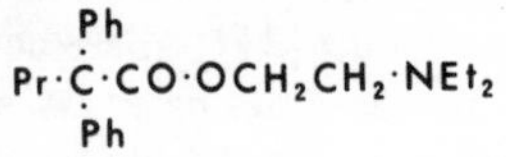

$$\mathrm{Pr{\cdot}\underset{Ph}{\overset{Ph}{C}}{\cdot}CO{\cdot}OCH_2CH_2{\cdot}NEt_2}$$

SKF 525-A
(*4.14*)

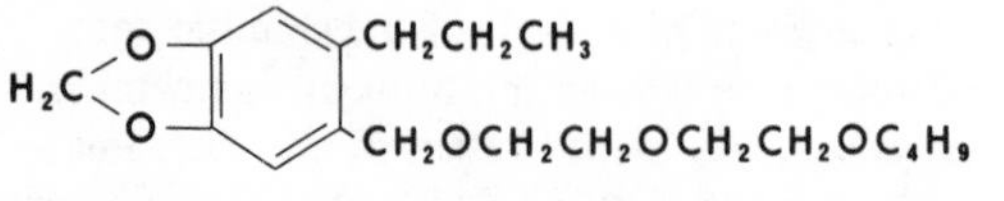

$$\mathrm{H_2C\langle O,O\rangle C_6H_2(CH_2CH_2CH_3)(CH_2OCH_2CH_2OCH_2CH_2OC_4H_9)}$$

Piperonyl butoxide
(*4.15*)

Apart from the synergism that arises from blocking sites of loss, examples of which were given in the foregoing, two other types are known. The first of these is sequential blocking (the inhibition of two or more consecutive metabolic processes) which is dealt with in Section 5.0. The other type of synergism is the attempt to retard the growth of bacterial mutants, based on observations that a mutant resistant to one drug does not easily undergo further mutation to a strain resistant to two or more drugs. It is for this reason that the weakly antitubercular substance *p*-aminosalicylic acid is included in the isoniazid therapy of tuberculosis (see Chapter 7 for discussion of drug-resistance).

Resembling synergism in its effects is the genetically determined lack (in an individual or a race) of a detoxifying enzyme, as a result of which a patient reacts to a small dose of a drug as though it were a large one. This phenomenon, which differs from gradual sensitization to a particular drug (an immune response, with its own standard symptoms quite unlike those of synergism), is termed pharmacogenetics.

The induction of drug-destroying enzymes. Examples of unintentional *over*dosage were described above; in these a drug blocks an enzyme that normally detoxifies a second drug taken at the same time (e.g. phenelzine taken with with pethidine). In contrast to this type of mishap, a patient may be subjected to *under*dosage through a drug-induced induction of *e.r.* enzymes [38]. The anti-rheumatic drug phenylbutazone is one of several drugs known to induce the excessive production of these enzymes, so that a fixed daily dose eventually produces an ever-decreasing effect; this is a consequence of the faster rate of destruction. Thus, 24 hours after a dose of 0·1 g/kg of phenylbutazone, the plasma of a dog showed a concentration of 100 μg/ml; but after five consecutive daily doses, the level of the drug had fallen to 15 μg/ml.

Similarly, successive doses of various barbiturates in mice and rats produced shorter and shorter periods of sleep. A similar induction of the barbiturate-destroying enzyme occurs in man within a week of the patient's beginning to take a small, nightly dose. If he continues to increase the dose, habituation and withdrawal symptoms are to be expected. On the other hand, the patient may discontinue the drug as soon as he feels the first decrease in action. After one or two weeks, the excess of enzyme will have disappeared and the original potency of the drug will be available.

That an inducing drug actually increases the amount of destructive enzyme in the *e.r.* has often been shown, e.g. by administering an azo-dye to laboratory animals for several days; then, when excretion of the dye was sharply diminished, hepatic *e.r.* was isolated, and the relevant enzyme found, by assay, to have increased greatly [39]. In one experiment in dogs, the amount of enzyme did not return to normal until 10 weeks had elasped.

Examples of other substances which stimulate their own metabolic destruction are chlorocyclizine, probenecid, tolbutamide, aminopyrine, meprobamate, glutethimide, chlorpromazine, chlordiazepoxide, methoxyflurane, 3, 4-benzpyrene, and DDT.

Moreover, heavy dosage with a drug can

induce increased production of an enzyme capable of destroying a different drug introduced into the dosage scheme simultaneously, or many days later. For example phenylbutazone, also barbiturates, speed up the metabolism of the coumarin anticoagulants in man. Hence a patient on anticoagulant therapy may be worse off if (as often happens) a barbiturate is also prescribed. For example a patient, who was taking 75 mg daily of bishydroxycoumarin and was later given 60 mg daily of phenobarbitone as well, showed a large decrease in plasma level of the coumarin drug, and the anticoagulant power fell. Yet soon after the phenobarbitone was discontinued, the coumarin drug regained its former level and so did the prothrombin time. Examples of pairs of drugs, the first of which can accelerate the metabolic destruction of the other *in man*, are: phenobarbitone and diphenylhydantion, phenobarbitone and griseofulvin, phenylbutazone and aminopyrine, and phenobarbitone and digitoxin. No less dangerous to the patient is the acceleration of destruction of the body's steroid hormones by various drugs such as phenobarbitone, chlorcyclizine, and phenylbutazone.

The primary site of antagonism of these inductive antagonists is the DNA core of RNA-polymerase. By increasing the activity of this enzyme, more RNA is synthesized, and finally more of the metabolizing enzymes [40].

An organism's normal reaction to a foreign substance is simply to burn it for food. Heteroaromatic nitrogen-containing nuclei have proved the most resistant to these destructive processes, and hence are being used more and more in designing new agents.

4.8 Conclusion

The foregoing discussion of selectivity arising through favourable differences in distribution inevitably led to consideration of the many factors by which distribution is controlled. To ensure the successful passage of a drug from the site of absorption to the receptor requires knowledge of the influences, good and bad, which it will meet on the way. To know what these are, and how they may be quantified are of the greatest importance in devising agents of improved selectivity, whether this selectivity depends primarily on distribution or on the phenomena described in Chapters 5 and 6. Undoubtedly that is why this is the longest chapter in the book!

References

[1] Lazarus, M and Rogers, W.P. (1951), *Aust. J. sci. Research, B*, **4**, 163–179.

[2] Franklin, T.J. (1963), *Biochem. J.*, **87**, 449–453; (1966), *Symp. Soc. gen. Microbiol.*, **16**, 192–212.

[3] Williams, A.C. and Klein, E. (1970), *Cancer*, **25**, 450–462.

[4] Kruger, P.G. (1955), *Radiation Research*, **3**, 1–17.

[5] Jensen, E.V. and Jacobsen, H.I. (1962), *Recent. Prog. Hormone Research*, **18**, 387–414

[6] Anderson, K.M. and Liao, S. (1968), *Nature (Lond.)*, **219**, 277–279

[7] Veldestra, H. (1956), *Pharmacol. Rev.*, **8**, 339–387.

[8] Goldstein, A., Aronow, L., and Kalman, S.M. (1974), *Principles of Drug Action*, Wiley, N.Y.

[9] Gaddum, J.H. (1936), *Proc. Roy. Soc. Med.*, **29**, 1373–1378.

[10] Nelson, E. and O'Reilly, I. (1960), *J. Pharmacol.*, **129**, 368–372.

[11] Wilbrandt, W. (1964), *Schweitz. med. Woch.*, **94**, 737–745.

[12] Krüger-Theimer, E. and Bünger, P. (1961), *Arzneimittel Forsch.*, **11**, 867–874.

[13] Collander, R. (1954), *Acta Chem. Scand.*, **5**, 774–780.

[14] Leo, A. and Hansch, C. (1971), *J. Org. Chem.*, **36**, 1539–1544.

[15] Fujita, T., Iwasa, J., and Hansch, C. (1964), *J. Amer. Chem. Soc.*, **86**, 5175–5180; Iwasa, J., Fujita, T., and Hansch, C. (1965), *J. Med. Chem.*, **8**, 150–153.

[16] Wilbrandt, W. (1959), *J. Pharm. Pharmacol.* **11**, 65–70.
[17] Martin, K., *Brit. J. Pharmacol.*, **36**, 458–469.
[18] Chappell, J. (1966), *Biochem. J.*, **100**, 43P.
[19] Barrett, E.G., Ralph, A., and Munday, K.A. (1970), *Biochem. J.*, **116**, 537–538.
[20] Bergel, F. (1958), *Ann. N.Y. Acad. Sci.*, **68**, 1238–1245.
[21] Schanker, L.S. (1961), *Ann. Rev. Pharmacol.*, **1**, 29–44.
[22] Albert, A. and Serjeant, E.P. (1971), *The Determination of Ionization Constants*, 2nd Edn., Chapman and Hall, London, 115 pp.
[23] Schanker, L.S., Shore, P.A., Brodie, B.B., and Hogben, C.A.M. (1957), *J. Pharmacol.*, **120**, 528–539.
[23a] Mitchell, P and Moyle, J. (1959), *J. Gen. Microbiol.*, **20**, 434–441.
[23b] Gale, E.F. (1947), *J. Gen. Microbiol.*, **1**, 53–76.
[24] Bittar, E. (ed.) (1970–1971), *Membranes and Ion Transport*, (3 vols.), Wiley, N.Y.
[25] Tatum. A.L. and Cooper, G.A. (1934), *J. Pharmacol.*, **50**, 198–215.
[26] Tréfoüel, J., Tréfoüel, Mme. J., Nitti, F., Bovet, D. (1935), *Comp. rend. Soc. Biol.*, **120**, 756–758.
[27] Welch, A.D. (1961), *Cancer Res.*, **21**, 1475–1490.
[28] Clemons, G.P. and Sisler, H.D. (1969), *Phytopathology*, **59**, 705–706.
[29] Wain, R.L. (1964), in *The Physiology and Biochemistry of Herbicides* (ed. Audus, L.) Academic Press, London.
[30] Rolinson, G.N. and Sutherland, R. (1965), *Brit. J. Pharmacol. Chemother.*, **25**, 638–650.
[31] Fouts, J.R. (1962), *Federat. Proc.* **21**, 1107–1111; Gillette, J.R. (1966), *Adv. Pharmacol.*, **4**, 219–261.
[32] Brodie, B.B. (1956), *J. Pharm. Pharmacol.*, **8**, 1–17.
[33] Brodie, B.B. (1971), *Chem. Biol. Interactions*, **3**, 247.
[34] Brodie, B.B. (1964), *The Pharmacologist*, **6**, 12–26.
[35] Dring, L.G., Smith, R.L., and Williams, R.T. (1970), *Biochem. J.*, **116**, 425–435.
[36] Fishman, W. (ed.) (1970-), *Metabolic Conjugation and Metabolic Hydrolysis* (appearing in several parts), Academic Press, N.Y.
[37] Casida, J.E. (1970), *J. Agr. Food. Chem.*, **18**, 753–772.
[38] Conney, A.H. and Burns, J.J. (1962), *Advances Pharmacol.*, **1**, 31–58.
[39] Porter, K.R. and Bruni, C. (1959), *Cancer Res.*, **19**, 997–1009.
[40] Gelboin, H.V., Wortham, J.S., and Wilson, R.G., (1971), *Nature, Lond.*, **214**, 281–283.

Suggestions for Further Reading

Albert, A. (1973), *Selective Toxicity* 5th edn., Chapman & Hall, London, 597 pp.

Covers the various aspects of the present chapter in greater depth, plus much related material.

LaDu, B., Mandel, H., and Way, E. (1971) Fundamentals of Drug Metabolism and Drug Disposition, Williams and Williams, Baltimore, 644 pp.

Useful extension material, particularly for Sections 4.2 and 4.5.

Brodie, B.B. and Gillette, J.R., *Concepts in Biochemical Pharmacology* (1971), Parts 1 and 2, Springer, Berlin, 487 and 772 pp.

These volumes form part of the larger series: *Handbuch der experimentellen Pharmakologie*; they provide many fresh ideas and examples of most of the topics discussed in the present chapter).

Levy, G. and Gibaldi, M. (1972), *Ann. Rev. Pharmacol.*, **12**, 85–98.

A valuable review of pharmacokinetics.

Bittar, E. (ed.) (1970–1971), *Membranes and Ion Transport* (3 vols), Wiley, N.Y.

Eisenberg, D. and Kauzmann, W. (1969), *The Structure and Properties of Water*, Clarendon Press, Oxford.

Rang, H.P. (ed.) (1974), *Drug Receptors*, Macmillan, London.

Proceedings of a 1972 symposium.

Davies, D.S. and Pritchard, B.N.C. (eds.) (1974), *Biological Effects of Drugs in Relation to their Plasma Concentrations*, Macmillan, London.

Parke, D.V. (1975), *Xenobiotics*, Chapman and Hall, London, 64 pp.

The fate of foreign substances in biological systems.

The monthly periodical *Xenobiotica* (Taylor and Francis, Ltd, London).

5 Favourable differences in biochemistry: the second principle of selectivity

5.1 Introduction

Of the three approaches to selectivity outlined in Chapter 2, this one has proved particularly successful in developing new and much used agents. In retrospect, it is astonishing how slowly comparative biochemistry developed as a subject of enquiry. Baldwin's book *An Introduction to Comparative Biochemistry* first appeared in 1937 and opened many eyes to the interest and possibilities of the subject. The last two decades have seen a steady growth of activity in comparative biochemistry, and today it is a thriving and successful subject.

As between species, few differences have been found in the way food is broken down and the resultant energy stored; but there are many differences in the ways in which essential metabolites are built up, and foreign substances broken down. Obviously enzymes, also their sub-substrates and coenzymes, which play a very active part in maintaining these differences, furnish useful sites on which selectivity can operate. Before discussing species differences in enzymes, it is useful to note that many differences have been found in the proportions and nature of enzymes in the different tissues of a *single organism*. Thus aconitase and oxaloacetate transacetase are much more abundant in heart than skeletal muscle, but the reverse is true of aldolase (see Table 5.1).

In mouse tumour cells, the amount of purine phosphoribosyl transferase activity was found to lie between 15 and 60 times the activity of that in normal cells from the same animal [1].

Table 5.1 Proportions of enzymes in muscle (rat) [2].

Enzyme	*Heart*	*Skeletal*
Aconitase	6	1
Aldolase	1	16

The more different the tissues of a mammal, the greater the difference in the kinds of enzymes maintained there. For example, the following three enzymes which liver and kidney possess in such abundance are absent from muscle: xanthine oxidase, catalase, and D-aminoacid oxidase. Moreover arginase occurs only in the liver, alkaline phosphatase only in the intestinal mucosa, 5-nucleotidase in the testis and α-mannosidase in the epididymis. The blood is disproportionately rich in carbonic anhydrase and the pancreas in ribonuclease. Glutamine synthetase, which condenses glutamic acid with ammonia, is abundant in brain and liver, but almost absent from all other human tissues [2]. Looking to the future, it is not hard to visualize a new race of pro-drugs, selectively converted to the true drug by a specific enzyme resident in the target organ.

5.2 Analogous enzymes

Many examples are known where enzymes carrying out apparently identical functions in two dissimilar organisms are themselves dissimilar. These are usually known as *analogous enzymes*. (They have also been called homologous enzymes. Isoenzymes, a sub-class of analogous enzymes, usually show only small differences).

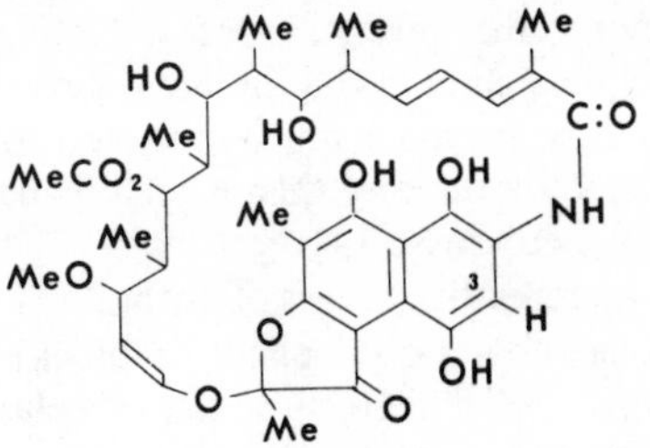

Analogous enzymes have been demonstrated in many ways: by differences in kinetics, electrophoresis, or specificity of coenzymes, substrates, or inhibitors.

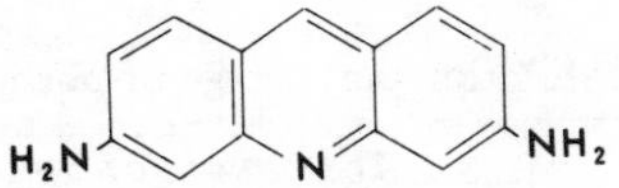

The first of the rifamycins, highly selective antibiotics, was isolated from *Streptomyces mediterranei.* This chemically-unstable prototype was converted in the laboratory to rifamycin SV (*5.1*) (usually called simply rifamycin), and there are variants, modified in the 3-position, with special names such as rifampicin. They all act by combining with the protein of bacterial RNA polymerase (DNA dependent) thus bringing all synthesis of RNA to a halt. They combine strongly, though not covalently, with this protein through its β-chain, a structure which does not occur in the RNA polymerases of higher organisms. Hence rifamycins exhibit selectivity in the highest degree, for they have no action at all on the RNA-synthesizing mechanism of mammals [3]. All details of how they act on the bacterial enzyme are not yet known, but the flat, aromatic ring in the drug and several of its free hydroxy groups are essential. Apart from the naphthalene ring, the molecule is far from flat, and must be visualized in a three-dimensional arrangement. In the clinic, where rifamycins are used for treating drug-resistant cases of tuberculosis and staphylococcal infections, rifamycin has been found free from toxicity to the patient.

Rifamycin
(*5.1*)

The simple aminoacridines [e.g. proflavine (*5.2*) (3, 6-diaminoacridine)] also, strongly inhibit RNA polymerase, but at a different site, namely by binding to the DNA starter (see p. 16) (they inhibit DNA polymerase by the same mechanism). Aminoacridines selectively bind to closed loops of DNA which are found in bacteria, viruses, and the mitochondria of higher organisms. These substances, particularly 9-aminoacridine, make excellent selective antibacterials for topical application, but lack the high selectivity of rifamycin for systemic use. However acridines with a basic side-chain, e.g. mepacrine ('Atebrin'), have proved to be excellent selective drugs for treating malaria, and so has chloroquin, a closely related quinoline which also inhibits RNA and DNA polymerases in the same way. At present, chloroquin is probably the most used drug for treating attacks of malaria and is considered sufficiently selective for this task, although some side-effects have been encountered.

Proflavine
(*5.2*)

Phenanthridine is an isomer of acridine. Aminophenanthridines also inhibit these two enzymes by binding to the DNA templates and are selective for the closed loop type of DNA. They are much used for the treatment of trypanosomiasis in cattle.

Schistosome worms provide an example where the selective action of a traditional remedy can be traced to the existence of analogous enzymes, one in the parasite, the other in the host. These worms are blood flukes (trematodes) parasitic in man, and causing a chronic, debilitating illness (see p. 9). As Table 5.2 shows, the phosphofructokinase of these worms (i.e.

the enzyme that converts fructose-6-phosphate to the diphosphate in the Embden-Meyerhof course of glycolysis) is much more sensitive to inhibition by antimony than is mammalian phosphofructokinase [4].

The therapeutic success of antimonial drugs in treating bilharziasis depends on blocking this enzyme selectively, although this class of drugs can cause unpleasant side-reactions. Because glycolysis is the main source of energy for this parasite, and this enzyme is a pacemaker, its substrate (fructose-6-phosphate) accumulates. Several other enzymes in these worms were shown to differ from their analogous mammalian types, though not so markedly as the above [4].

Dihydrofolic acid
(*5.3*)

Table 5.2 Inhibition percentage of phosphofructokinase by antimonial drugs [4].

Concn. of antimonial M × 10^5	*Antimony potassium tartrate* (A) Enzyme from *Schistosoma mansoni*	(B) Enzyme from rat brain	*Stibophen* as (A)	as (B)
100	100	32	100	0
50	100	4	100	0
30	100	0	85	0
10	70	0	44	0
3	32	0	0	0
1	2	0	0	0

5.3 More analogous enzymes: The dihydrofolate reductases

The biosynthesis of dihydrofolic acid (*5.3*), from glutamic acid, *p*-aminobenzoic acid, and a substituted pteridine, was referred to on p. 14.

As a first step in converting dihydrofolic acid into a series of coenzymes, the enzyme dihydrofolate reductase converts it to tetrahydrofolic acid, by reducing the 5, 6-double bond. The insertion of various one-carbon substituents then gives the appropriate coenzyme. For example, the coenzyme for the insertion of C-2 into the purine nucleus, i.e. for the formylation of the imidazole (*5.4*) to give inosinic acid (*5.5*), is $N_{(10)}$-formyl-5, 6, 7, 8-tetrahydrofolic acid. Similarly the coenzyme for the insertion of C-2 into the imidazole (*5.4*) is $N_{(5)}$, $N_{(10)}$-methenyl-5, 6, 7, 8-tetrahydrofolic acid. It is evident that the C-2 in

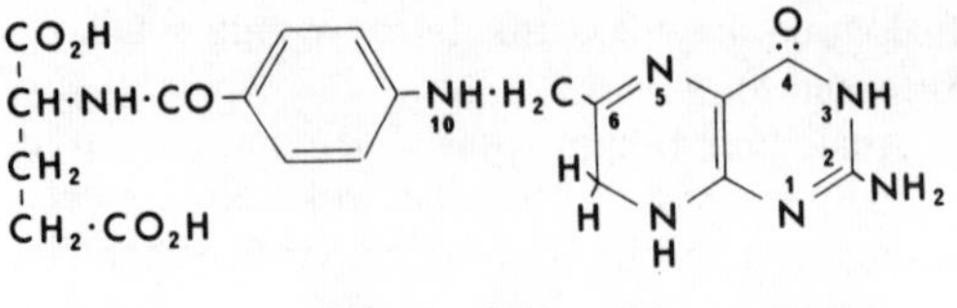

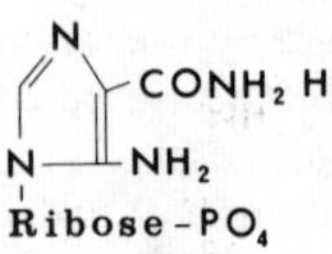

'AICR' (*5.4*) | Inosinic acid (*5.5*) | Thymine (*5.6*)

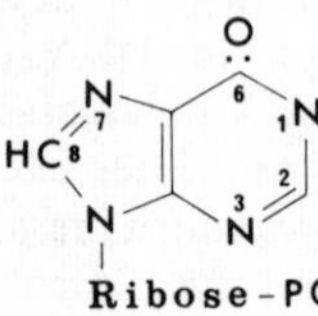

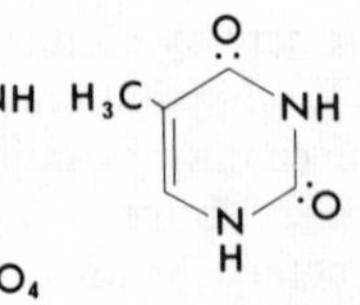

(*5.4*) becomes C-8 in the purine (*5.5*). Because all biosynthesis of purines follows this pathway, any lack of tetrahydrofolic acid must cause a gross purine deficiency in the cell.

A related coenzyme, $N_{(5)}$, $N_{(10)}$-methylene-5, 6, 7, 8-tetrahydrofolic acid is responsible for inserting the 5-methyl group into uridylic acid to give thymidylic acid which is essential for DNA formation and cannot be biosynthesized in any other way. In many organisms, the first biochemical lesion after dihydrofolate reductase has been inhibited is lack of thymidylic acid. Unless thymine (*5.6*) is supplied in some form, death soon takes place. In some other organisms, the purine deficiency proves more lethal.

The first drug found to block dihydrofolate reductase was aminopterin, quite simply made by changing the 4-oxo group of folic acid to a primary amino-group. Clinical trials in cancer wards, from 1958 onwards, showed that methotrexate (amethopterin) was a more selective drug, and this quickly replaced aminopterin. Methotrexate differs from dihydrofolic acid (*5.3*) in two details: the replacement of the oxo by an amino-group as in aminopterin, and the replacement of the 10-hydrogen atom by a methyl group. Methotrexate is in regular use for treating the leukaemia of young adults (acute lymphatic leukaemia) [5]. It brings about remission of symptoms, but eventually the leukaemic cells develop resistance to the drug by increasing the production of dihydrofolate reductase (see Chapter 7 for discussion of resistance). This undesired effect is now prevented by synergistic therapy, namely the simultaneous administration of 6-mercaptopurine and a corticosteroid [6].

In two other types of cancer, methotrexate brings about a lasting cure. Choriocarcinoma, a fast-growing tumour of pregnancy with normally a high death-rate, is quickly and completely cured by methotrexate. This drug, too nearly always effects a complete cure of Burkitt's lymphoma which is a highly malignant cancer affecting African children living in hot, wet areas. It usually begins in the jaw, spreads throughout the body, and kills the patient within six months. It must be pointed out that methotrexate has toxic side-effects to humans and hence is most suited for treating those malignancies that yield rapidly to it.

Altered folic acids, such as methotrexate, have little toxicity for most bacteria and protozoa because they are not taken up readily by these organisms [7]. About 1950, G.H. Hitchings initiated a programme to simplify the pteridine part (found to be the active portion) of aminopterin in order to achieve better penetration. The first success followed omission of the nitrogen atom from position five, giving simple 2, 4-diamino-1, 3, 8-triazanaphthalenes. Greater success followed further simplification and it was soon found that 2, 4-diaminopyrimidines have a powerful anti-folic action on microorganisms. This knowledge was used at once to develop the powerful antimalarial drug, pyrimethamine (*5.7*) [8].

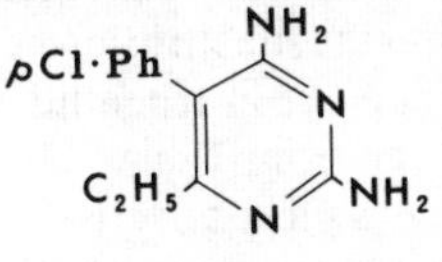

NH₂ | OMe | NH₂
pCl·Ph | CH₂ | N
C₂H₅ N NH₂ | MeO MeO | N NH₂

Pyrimethamine (*5.7*) | Trimethoprim (*5.8*)

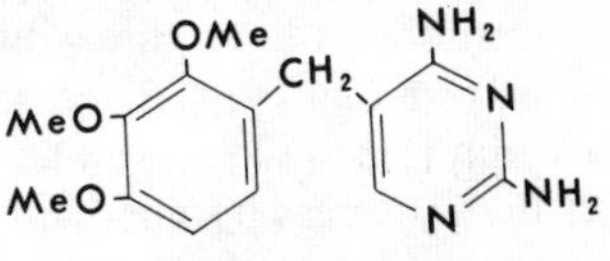

Pyrimethamine (2, 4-diamino-6-ethyl-5, *p*-chlorophenylpyrimidine) has become the most widely used of all prophylactics against malaria. The lipoidal groups in the molecule (*5.7*) favour its uptake by the tissues containing the malarial parasite, and they also increase the adsorption of the drug on the dihydrofolate reductase by van der Waals forces [9]. Dihydrofolate reductase isolated from a malarial parasite *Plasmodium berghei*, was found to have a molecular weight of 200 000, which is 10 times as large as those of analogous enzymes purified from mammals and bacteria.

Table 5.3 Concentrations ($\times\ 10^8$ M) of antifolic drugs needed for 50 per cent inhibition of dihydrofolate reductase, isolated from 6 sources [10], [11], [12].

Substance	*Human liver*	*Rat liver*	*Mouse erythrocyte*	*Pl. Berghei*	*Tryp. equiperdum*	*E. coli*
Pyrimethamine (*5.7*)	180	70	100	0·05	20	2500
Trimethoprim (*5.8*)	30 000	26 000	100 000	7·0	100	0·5
Methotrexate	9	0·2	(not done)	0·07	0·02	0·1

As Table 5.3 shows, the plasmodial enzyme is inhibited by pyrimethamine at a concentration about 2000 times lower than that inhibiting the analogous mammalian enzymes. The concentration that inhibited the plasmodial enzyme corresponded to that achieved in the tissues after the usual prophylactic dose. These data established that the selective action of pyrimethamine in malaria is due to the extraordinary sensitivity of the enzyme in the parasite compared to that in the host [10].

The inhibition of isolated dihydrofolate reductase by pyrimethamine was not greatly altered when the 5-phenyl- was changed to a butyl-group [9]. For effective antibacterial action in mammals, the 2, 4-diaminopyrimidines were found to require a different substitution pattern of alkyl and aryl groups, as in trimethoprim (*5.8*) [13]. It can be seen from Table 5.3 that, whereas pyrimethamine is most selective against the malarial parasite enzyme, trimethoprim is selective against a bacterial enzyme as well. Compared with the high selectivity shown by these diaminopyrimidines, the pteridine (methotrexate) exhibits a much narrower range.

Table 5.4 exemplifies the high selectivity that trimethoprim exerts against the dihydrofolate reductase of bacteria, both Gram-positive and -negative types, while leaving the analogous mammalian enzyme unharmed.

It is noteworthy that these useful 2, 4-diaminopyrimidine drugs have been discovered mainly by the exercise of scientific reasoning. Maximal activity was achieved when the basic pK_a value lay between 6 and 8, an indication that cell membranes have to be traversed (see p. 19).

The therapeutic value of trimethoprim has been increased by using it in conjunction with a sulphonamide (sulphamethoxazole) (*5.9*) for sequential blocking, a technique that will now be explained.

Table 5.4 Effect of trimethoprim (*5.8*) on isolated dihydrofolate reductase [10]. Concentration ($\times\ 10^8$ M) causing 50% inhibition

SOURCE:	*Mammalian liver*		*Bacteria*	
	Man	30 000	*E. coli*	0·5
	Rat	26 000	*S. aureus*	1·5
	Rabbit	37 000	*P. vulgaris*	0·4

Current biochemical studies reveal that growth-factors are gradually built up from components moved along the enzymatic equivalent of a factory's production line, each stage of assembly being carried out by a different enzyme. The arithmetic of sequential blocking is this: if the first enzyme is blocked to the extent of 90 per cent, then only 10 per cent of the partly completed factor reaches the second enzyme. If one is fortunate enough to discover how to block the second enzyme also by 90 per cent, then only 1 per cent of the partly completed factor emerges, and that may be too little to sustain life in the parasite. It may be asked why it is not sufficient to use more of the first drug and block the first enzyme by 100 per cent. The answer is: because the usual shape of a dose-response curve is hyperbolic, increasing the concentration of a drug beyond a 90 per cent inhibition seldom leads to a worthwhile extra response, and it would usually call for a higher drug dosage than is compatible with the patients welfare.

In the sequence under consideration,

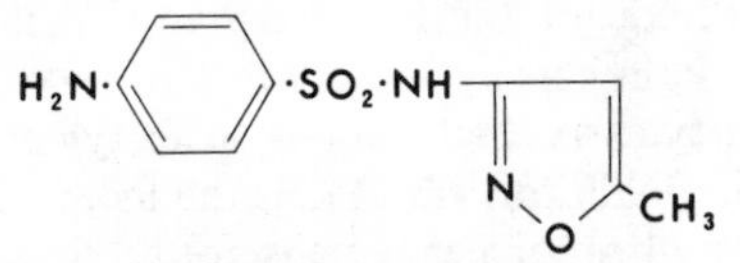

$H_2N \cdot C_6H_4 \cdot SO_2 \cdot NH$–(5-methylisoxazol-3-yl)

Sulphamethoxazole
(*5.9*)

sulphamethoxazole blocks the incorporation of *p*-aminobenzoic acid into dihydrofolic acid, and trimethoprim prevents the reduction of this pteridine to tetrahydrofolic acid. This combination is official in the British Pharmacopoeia under the name Co-trimoxazole (Brand names: 'Septrin', 'Bactrim', 'Eusaprim'). It has been found extraordinarily effective in bacterial dysentery, bronchitis, and long-standing infections of the urinary tract whether with Gram-positive or -negative organisms [14]. Combinations of 2, 4-diaminopyrimidines and sulphonamides have also been found useful in malaria and other protozoal diseases [15]. Prolonged medication with anti-folic drugs (but not with sulphonamides) has an adverse effect on red blood cells, leading to macrocytic anaemia. This side-effect may be prevented or overcome quite readily be administering folic acid or, better, $N_{(5)}$-formyl-tetrahydrofolic acid, without diminishing the effect of the anti-folic drug on the uneconomic organism which cannot absorb the antidote [16]. More serious, the chance of abortion or damage to the foetus makes it necessary to use anti-folic drugs cautiously during pregnancy.

5.4 Absence of enzymes from one of the species.

I. The sulphonamides. The organophosphates. The basis of the selectivity of the antibacterial sulphonamides, outlined in Chapter 2, depends on two reinforcing deficiencies. (i) Mammals lack the enzymes needed for the synthesis of dihydrofolic acid, and hence they are tolerant of these sulphonamides, which act on bacteria by interfering with this synthesis (see p. 14). (ii) Bacteria lack the permease with the aid of which mammals absorb dihydrofolic acid from the diet. With the help of data already given (pp. 11 and 14) a clear account can be constructed, and need not be enlarged upon here.

A similar situation will now be described for the organophosphate insecticides where selectivity has been reinforced by making use of the absence of two enzyme processes, one in the economic and one in the uneconomic species.

The toxicity of organic phosphorus compounds is directly proportional to the inhibition of the enzyme acetylcholinesterase. When the functioning of this enzyme is suspended, acetylcholine accumulates at nerve-muscle junctions in vertebrates, and at nerve-nerve junctions (synapses) in both insects and vertebrates. Although a natural neurotransmitter, the presence of acetylcholine in excess leads to incoordination and (in greater excess) death.

This class of substrates was developed in Germany, first as chemical defence agents ('nerve gases') during the Second World War, and after that as agricultural insecticides through the pioneer efforts of G. Schrader. The first organophosphate insecticides used in the field were insufficiently selective, and there were many casualties among birds, and even men. Before relating how these deficiencies were overcome, some of the underlying biochemistry must be reviewed.

The phosphorus insecticides block their target enzyme (receptor) by phosphorylating it covalently. For example paraoxon (*5.10*) diethylphosphorylates the enzyme and liberates *p*-nitrophenol. The site of phosphorylation is quite specifically the hydroxy group of serine [17]. Electron-withdrawing groups, which weaken the P-O bond, increase the activity up to a maximum, beyond which any increase in electron-affinity leads to wasteful hydrolysis by water. Below this maximum, there is a linear relationship between (a) *log* hydrolysis constant and (b) the *log* bimolecular rate constant for the reaction between the enzyme and the organic phosphate. The strong β-emission of ^{32}P, and its long half-life of 14 days, makes isotopic phosphorus a very convenient marker for following the behaviour of the phosphorus insecticides in tissues.

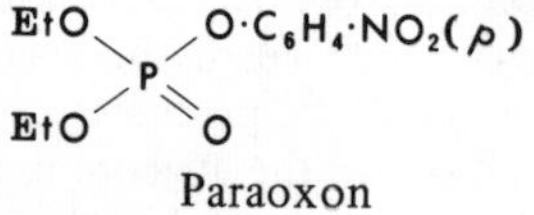

$$(EtO)_2P(=O)\cdot O\cdot C_6H_4\cdot NO_2(p)$$

Paraoxon
(*5.10*)

The inactivated enzyme can be reactivated by oximes, if prompt action is taken. What happens is that the inhibited enzyme $(EtO)_2P(:O)\cdot OR$ reacts with the anion of the antidote, namely $X_2C:NO^-$ to give the reactivated enzyme HOR plus the by-product $(EtO)_2P(:O)\cdot ON:CX_2$. If the antidote is not administered promptly, the inhibited enzyme 'ages' by hydrolysis of an ether group to give $-O(EtO)P(:O)\cdot OR$. Because of Coulombic repulsion by the two anions, the antidote cannot now approach the inhibited enzyme and the pathological state of the affected organism (whether insect or man) cannot now be reversed.

A parenthetical note on analogous enzymes is inserted here, before continuing with the main story. Acetylcholinesterases isolated from various insect species differ greatly in their susceptibility to phosphorus insecticides [18]. The acetylcholinesterase in the worm *Haemonchus contortus*, which parasitizes the sheep's gut, is irreversibly inhibited by the organophosphorus drug haloxon (see below). Yet the acetylcholinesterase of the sheep's gut is only temporarily affected and rapidly recovers. Other species of worm, not affected by haloxon, have been shown not to carry this *Haemonchus* variant of the enzyme [19].

After a limited success in 1944 with parathion which only the insect can convert to the toxicant paraoxon (*5.10*), the highly selective malathion (*5.11*) was introduced in 1950. This substance, (*S*-[1, 2-di(ethoxycarbonyl)ethyl] *O*, *O*-dimethylphosphorodithioate) is not the actual toxicant, but is hydrolysed to malaoxon (*5.12*) in insects by an appropriate enzyme that they carry. Malaoxon is an acetylcholinesterase inhibitor, and it kills the insect. The P = S hydrolysing enzyme is not present in vertebrates which go comparatively unharmed. Simultaneously, mammalian (non-specific) esterases hydrolyse the carboxylic ester (CO_2Et) group in malathion to a carboxylic acid (CO_2H) group, a type of structure which the kidney rapidly eliminates. Insects are very poorly equipped to carry out hydrolyses of esters. Thus high selectivity is achieved by two metabolic changes, one of them toxifying and one detoxifying, depending on the organism [20].

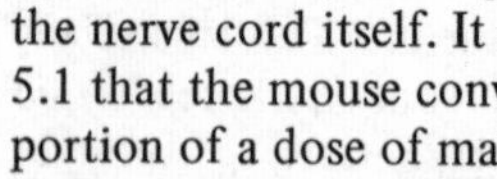

$$(MeO)_2P(=S)\cdot S-CH(CO_2Et)\cdot CH_2\cdot CO_2Et \qquad (MeO)_2P(=O)\cdot S-CH(CO_2Et)\cdot CH_2\cdot CO_2Et$$

Malathion (*5.11*)   Malaoxon (*5.12*)

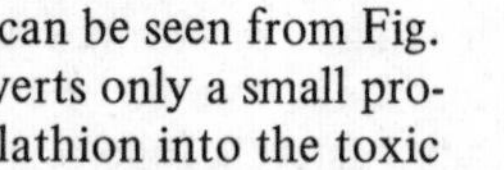

The P:S to P:O change in insects is effected by microsomes, in the gut, the fat body, and the nerve cord itself. It can be seen from Fig. 5.1 that the mouse converts only a small proportion of a dose of malathion into the toxic malaoxon, and eliminates the toxicant quickly, whereas the cockroach not only converts much more of the dose to the toxicant but retains it longer.

Similarly, Table 5.5 shows that the mouse can withstand a high dose of malathion because it does not readily convert it to the toxic derivative malaoxon, whereas the cockroach, and particularly the housefly, are excellent converters.

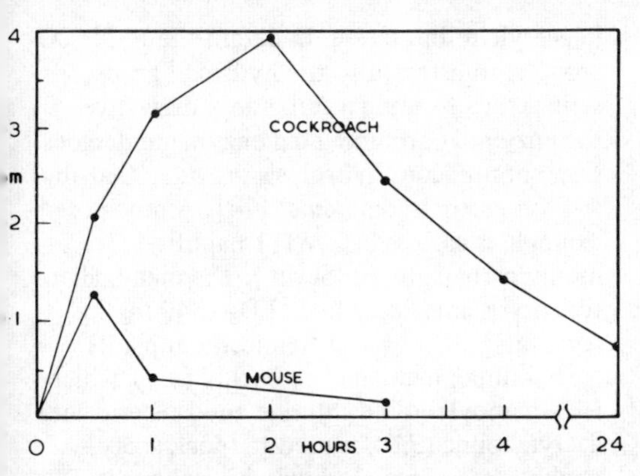

Fig. 5.1 Malaoxon level after injecting 30 ppm of ^{32}P-malathion [20].

Table 5.5 Toxicity of phosphorus insecticides to mammals and insects [20].
(LD_{50} in ppm w/w)

Species	*Malathion*	*Malaoxon*
Mouse (i/p)	1590	75
House fly	30	15
Cockroach	120	15

Mammalian carboxyamidases, too, readily hydrolyse amides to acids, but insects are almost unable to do so [21]. This feature is incorporated in the *systemic insecticide* dimethoate (*5.13*). Such systemic agents are absorbed by the plants which, containing no acetylcholinesterase, are unharmed; sucking and biting insects, however, are killed. This ingenious approach, which is much used, spares the harmless insects, some of which are needed for pollination, whereas others are the normal food of field birds.

Another group useful for ensuring detoxication by vertebrates is a methyl bearing a strong electron-attracting substituent to favour a metabolic break in the O-C linkage. Heterocycles with two or more doubly-bound nitrogen atoms exert the required electron depletion of this linkage. An outstanding insecticide of this type is diazinon (*5.14*). This agent, discovered in

MeO, MeO, P, S, S·CH₂·C(O)·NHMe

Dimethoate
(*5.13*)

EtO, EtO, P, S, O, N, N, CHMe₂, Me

Diazinon
(*5.14*)

1952, is *O, O'*-diethyl *O"*(2-isopropyl-4-methyl-6-pyrimidinyl) phosphorothioate [21]. It has proved extraordinarily successful, when sprayed on the hindquarters of sheep, in ridding them of the intense irritation and consequent loss of condition caused by the maggots of blowflies.

All these types of vertebrate detoxification have been accompanied by some hydrolysis of the ester groups attached to the phosphorus atom, giving substituted phosphoric acids which are quickly excreted by the mammalian kidney. This observation has led to development of a family of phosphorus insecticides where the toxic P = O linkage is pre-formed, but which achieve vertebrate detoxification by phosphate ester hydrolysis accelerated by neighbouring electron-attracting substituents. Such a substance is dichlorvos (*5.15*) a highly volatile fly-killer much used in homes and restuarants. Chemically it is dimethyl 2, 2-dichlorovinylphosphate. When inhaled, or absorbed through the skin, it is rapidly hydrolysed to dichlorovinylphosphoric acid which is harmless [22]. To insects, however, its toxicity is cumulative. Because of public concern, the American Conference of Government Industrial Hygienists made a special investigation of this product, and set a safe human threshold limit value, for dichlorvos in air, of 1 μg per litre [23]. 'Vapona' slow-release strips, when used as directed (one strip per 1000 ft^3), give off (for the major part of the life of the strip) a concentration about one-tenth of this.

Of the newer types of organophosphate

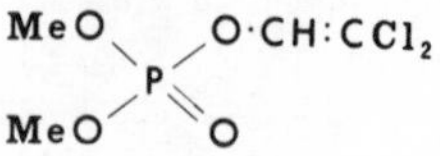

Dichlorvos
(*5.15*)

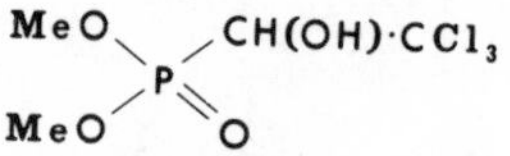

Trichlorophon
(*5.16*)

insecticides, some have been made so selective that they may be taken by mouth. Thus trichlorophon (*5.16*) is given orally, as a single dose, to cattle to provide durable protection against the hide-piercing warble fly. This substance, also known as metriphonate or 'Dyvon' is chemically dimethyl 2, 2, 2-trichloro-1-hydroxyethylphosphonate.

Worms do not degrade certain kinds of phosphorus insecticides. Thus dichlorvos (*5.15*) has been found a safe and effective oral vermifuge for dogs, horses, and swine. Haloxon (*5.17*), another orally administered anthelmintic, selectively kills nematode worms in sheep and cattle, oxyurid and ascarid worms in horses, and several species of worms in pigs and poultry. Chemically it is di-(2-chloroethyl) 3-chloro-4-methylcoumarin-7-ylphosphate. Insects as well as mammals can detoxify compounds containing the chloroethyl group present in haloxon, pointing to a new pattern of selectivity.

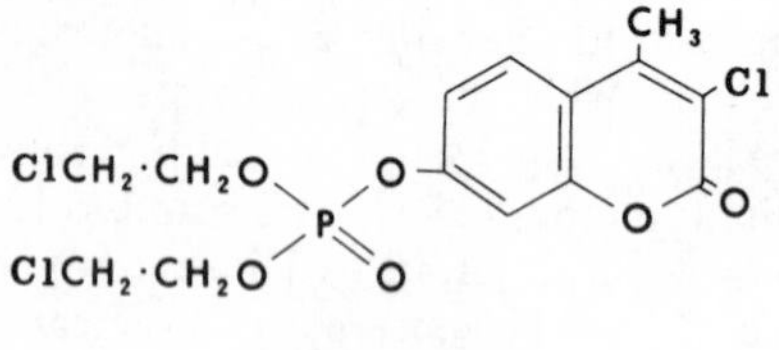

Haloxon
(*5.17*)

Carbamates (urethanes). From 1947 onwards, carbamate insecticides have been preferred for some uses, on the grounds that they are less toxic to humans and more selective among insect species than phosphorus insecticides. Like the latter, they act by acylating serine residues in acetylcholinesterase. In detail: the $R_2N \cdot CO$-group is transferred to the hydroxyl group of serine, thus forming a carbamoyl derivative of the enzyme. Carbamoyated enzyme undergoes slow spontaneous hydrolysis provided that the *N*-alkyl group is kept small [24]. A much used example is carbaryl (*5.18*) (1-naphthyl *N*-methylcarbamate, or 'Sevin'). The mammalian toxicity is amazingly low [LD_{50} 540 mg/kg (oral rat)]. It is used systemically in plants.

In human medicine, carbachol (*3.5*), which is carbamoylcholine chloride and isosteric with acetylcholine (*3.6*), is used to block acetylcholinesterase, which slowly hydrolyses it. It exhibits both the muscarinic and nicotinic actions of acetylcholine in a more prolonged and intense form. It is used to abolish postoperative atony of bowel and bladder; also to lower intraocular tension in glaucoma.

5.5 Absence of enzymes. II. Further examples. Striking differences in biochemistry are not often found between a malignant tumour and its tissue of origin. Nevertheless several cancers lack an enzyme, present abundantly in the latter. Thus the use of the purine analogue 8-azaguanine (*5.19*) in treating solid tumours, depends on the presence of a high content of guanase in the brain, liver, and kidney which destroys this drug, whereas common tumours

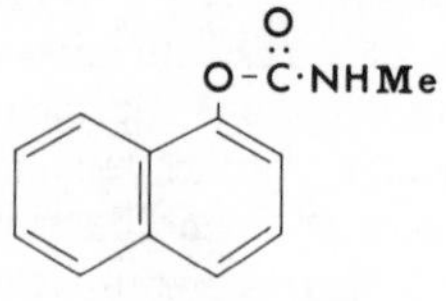

Carbaryl
(*5.18*)

(both primary and metastatic) in these organs lack this enzyme [25]. In patients with neoplasma in the head, for whom all other possible treatments have been exhausted, 8-azaguanine produces remissions.

Some strains of human leukaemia cells lack the enzymes for synthesis of L-asparagine, although this is necessary for their metabolism and growth. By injecting asparaginase into a patient, his peripheral body-pool of this aminoacid can be depleted without causing him great distress. In the cancer wards, drug-resistant lymphoblastic leukaemia of children respond to this treatment, but for a short time only [26].

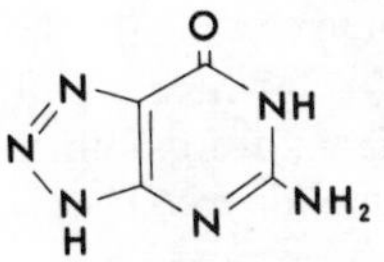

8-Azaguanine
(*5.19*)

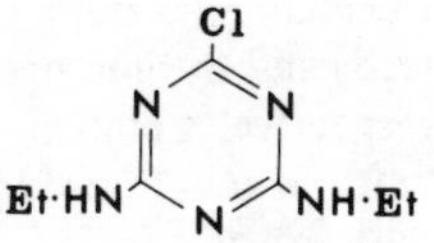

Simazine
(*5.20*)

Pyrethrins are rather expensive insecticides, with a high degree of selectivity which enhances their value for domestic use. Chemically they consist essentially of chrysanthemic acid (a cyclopropane derivative) which is esterfied with a lipophilic alcohol. Apart from the natural pyrethrins (extracted from flowers) one can buy totally synthetic examples with improved properties. Mammals detoxify pyrethrins very quickly with non-specific oxidases. Normal insects lack such enzymes, but resistant strains have the equivalent [27].

Rotenone, an insecticide of vegetable origin and chemically an oxygen-heterocycle, blocks the dehydrogenation of NADH in the respiratory chain of all mitochondria at a dilution of 10^{-8} M. It thus prevents the oxidation of pyruvate and glutamate. Fish as well as insects are highly susceptible to rotenone, but mammals are rich in enzymes that rapidly metabolise it although *isolated* mitochondria are very susceptible [27].

Good selectivity is shown by the soil fungicide sodium *p*-dimethylaminobenzenediazosulphonate ('Dexon') which inhibits mitochondrial oxidation of NADH in the fungus *Pythium ultimum*. Sugar beets, which this fungus infects, have an enzyme in the mitochondria which decomposes this fungicide [28].

Simazine (*5.20*), one of the triazine herbicides which kill weeds by inhibiting photosynthesis, does no harm to cereal plants which possess a detoxifying enzyme capable of hydrolysing the chlorine substituent [29].

Some examples will now be given of clear biochemical differences which point to significant enzymatic differences between species but which have not yet been used to provide selective agents.

The aromatic aminoacids, phenylalanine and tryptophan, are made in bacteria and plants from shikimic acid (*5.21*). Mammals lack the enzymes for this pathway and hence have to obtain these aminoacids from their diet [30]. From shikimic acid, plants make gallotannins, and insects (with the help of resident bacteria) synthesize protocatechuic acid (3, 4-dihydroxybenzoic acid) needed to tan their integument. In bacteria, shikimic acid is the progenitor also of ubiquinone, vitamin K, and *p*-aminobenzoic acid (and hence of folic acid). Because shikimic acid does not enter into mammalian metabolism, its synthesis and use are clear targets at which to aim for purposes of selective toxicity. In planning such a programme, it must be borne in mind that intestinal bacteria provide man with his principal source of vitamin K, essential for the control of bleeding. Insect muscle utilizes the Meyerhof sequence only as far as

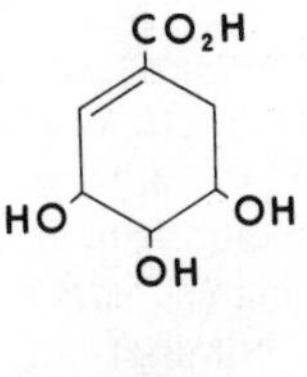

Shikimic acid
(*5.21*)

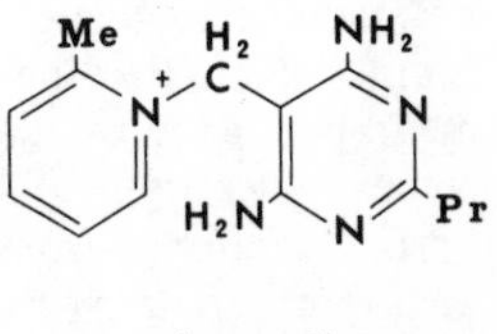

Amprolium
(*5.22*)

pyruvate, and the NADH produced during triosephosphate oxidation seems to be reoxidized by the reduction of dihydroxyacetone phosphate. The major sugar in the plasma is trehalose (a disaccharide of glucose) which plays a major part in the glucose transport system of insects (of worms, also). See Ref. [31] for a review of insect biochemistry, and Ref. [32] for one of helminth biochemistry. Flat worms cannot synthesize the fatty acids that they require for their triglycerides and phospholipids, but they have enzymes which alter the host's lipids to suit their requirements.

Although adenosine triphosphate is used by all forms of life for energy storage and transfer, it is assisted by phosphagens that differ between species. Phosphocreatine, which is the sole phosphagen for vertebrates, occurs in only a few invertebrate species, but phosphoarginine is more common. Some rarer phosphagens, with a distribution restricted to a few invertebrates, are phosphoguanidoacetic acid, phosphoguanidotaurine, and lombricine which is a derivative of D-serine. Many bacteria, quite astonishingly, accumulate β-hydroxybutyric acid as an energy reserve.

Micro-organisms often excrete chelating agents to combine with essential trace metals in the medium. Thus *E. coli* releases enterochelin (a cyclic trimer of *N*-2, 3-dihydroxybenzoylglycine) in order to chelate ferric iron in its environment. The resulting complex, in which iron is exceeding strongly bound, is absorbed selectively by this bacterial species, which has a hydrolytic enzyme to liberate the iron [33].

5.6 Metabolite analogues (mostly antagonists)

It is unfortunate that the word 'metabolite' has come to be used in two senses: (a) substrates and coenzymes essential in metabolism, and (b) products formed from foreign substances in an attempt to detoxify and eliminate them. The idea of metabolite analogues refers to the first sense. These analogues are drugs or other foreign substances especially designed to antagonise metabolites. The molecule of each such analogue has a region similar to that region of the metabolite which makes contact with the enzyme. To be effective, this must be a similarity not only in dimensions, but also in electron distribution, because most of the active sites on enzymes are highly polar. Each analogue exerts its antagonism by occupying and blocking the enzyme site used by the metabolite. This subject has already been touched on in discussing the analogues sulphanilamide (*4.7*) carbachol (*3.5*) and pyrimethamine (*5.7*) which inhibit, respectively, enzymes designed for *p*-aminobenzoic acid (*3.8*), acetylcholine (*3.6*), and dihydrofolic acid (*5.3*). It is proposed here to expand the topic a little.

Some antagonists are of such a chemically simple nature that their relevance as metabolite analogues is easily overlooked. For example, inorganic cations are in competition with other inorganic cations.

Small molecules are sometimes antagonized by their nearest homologues, as succinic acid is by malonic acid. Common ways to obtain an antagonistic analogue are to replace one atom in a ring by another, or to exchange electron-attracting groups, e.g. $-CO_2H$ may be altered to $-COCH_3$, $-SO_2NH_2$, or $-SO_2OH$, taking care not to ionize any weakly basic group present. Sometimes hydrogen is successfully replaced by fluorine, or methyl by chlorine, but the alteration of hydrogen (radius 1·2 Å) to methyl (2·0 Å) is usually found to be too much for the enzyme to accept. The important criteria are that the analogue must be so similar to the natural metabolite that the enzyme is deceived into taking up the foreign molecule in place of its normal substrate or coenzyme. Yet the analogue must be dissimilar enough to be capable of functioning as the substrate does.

The relationship between metabolite and analogue is usually competitive. This is, if x molecules of metabolite are antagonized by *y* molecules of analogue, then 10*x* molecules of metabolite require 10*y* molecules of analogue to give the same biological endpoint, and so on. Because such competitive reactions are freely reversible, the antagonism of *x* molecules of metabolite by *y* molecules of analogue can be abolished by another *x* molecules of metabolite, and so on.

Each metabolite + antagonist pair has an unique *index of inhibition*, defined as the ratio of the number of molecules of analogue (to those of metabolite) required to give 50 per cent inhibition . This ratio, which varies with the biological species, expresses the relative affinity of analogue and metabolite for the appropriate receptor, but it also includes a hidden term for differences in the penetrative ability of the two substrates when the site of action is not exposed in the test. The dissociation constant (K_i) for an inhibitor is expressed:

$$K_i = [E]\,[I]/[EI],$$

where [E] is the concentration of the enzyme, [I] is that of the inhibitor, and [EI] is that of the complex which they form. Hence the index of inhibition (or inhibitory index) is K_i/K_m, in which K_m is the Michaelis-Menton constant:

$$K_m = [E]\,[S]/[ES]$$

where [S] is the concentration of the substrate. The smaller the index of inhibition, the more efficient the inhibitor. That for the antagonism by sulphanilamide of *p*-aminobenzoic acid in streptococci is 300, a rather large but still quite advantageous figure. The index 0·0001 for the antagonism by methotrexate of dihydrofolic acid (on its reductase) is outstandingly good.

Most inhibitory indices have been found to be high, and this is not surprising because enzymes have evolved, slowly and under the strong selection pressure of Nature, to handle their work efficiently. They are usually undersaturated with substrate and, although adsorption of the antagonist may initially inhibit the enzyme, accumulation of extra substrate can overcome this inhibition, largely leading to a new series of steady states, the first few of which may not disadvantage the organism greatly. This condition exemplifies the difference between simplified model systems studied in the laboratory and the similar, but more complex, systems in the living cell whose self-perpetuating properties are the very essence of life. These considerations point to the value of attacking what Krebs called the pacemaker enzymes [34] which are specially vulnerable to inhibition.

Although it is not difficult to make metabolite analogues, it has proved hard to find ones that are selective. This has occurred through failure to choose a metabolite that is important in the uneconomic species and yet unimportant in the economic species. Otherwise success must depend on the analogue being taken up more favourably by the uneconomic species. (Chapter 4 shows how this may be contrived). Examples will now be given of metabolite analogues that have surmounted all these hurdles and are in commercial production and regular use.

Analogues of vitamins will first be mentioned. Amprolium (*5.22*) ('Amprol'), an analogue of thiamine, has proved highly successful in coccidiosis, a protozoan infection of poultry. It is thought to inhibit the enzyme thiamine phosphorylase in the cytoplasmic membrane of the parasite [35]. Thiamine reverses its action.

Sodium 2, 2-dichloropropionate (dalapon, 'Dowpon') is much used for killing grasses in dicotyledonous crops such as beet and lucerne. It antagonizes the incorporation of pantoic acid into pantothenic acid. The latter reverses its action.

Among antibiotics, oxamycin (*6.2*) is an analogue of D-alanine (see p. 60), penicillin of certain dipeptides (p. 61), and chloramphenicol possibly of uridine (p. 60). The lethal action of oxamycin on bacteria can be prevented by D-alanine.

The antibiotic azaserine (*o*-diazoacetyl-L-serine), which is used as a sequential blocking agent in the chemotherapy of cancer, has been shown to act as a structural analogue of glutamine, and hence interferes with one of the earliest stages of purine biosynthesis [36].

Some successful analogues of purines and pyrimidines have already been mentioned: 8-azaguanine (*5.19*) (p. 51) and 5-fluorouracil (*4.1*) (p. 22). The latter is converted by the cell to fluorodeoxyuridylic acid which has an affinity for thymidylate synthetase several thousand times greater than that of deoxyuridylic acid. It is able to keep the substrate off the enzyme and no DNA can be synthesized [37]. 6-Azauracil (*5.23*), introduced for the treatment of tumours in mammals, is used in agriculture to inhibit powdery mildew. This substance is converted by the cell to the ribotide which blocks orotidylic decarboxylase and hence the last step in the biosynthesis of uridylic acid.

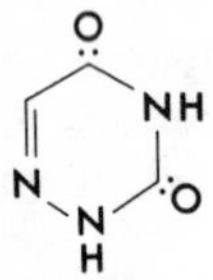

6-Azauracil
(*5.23*)

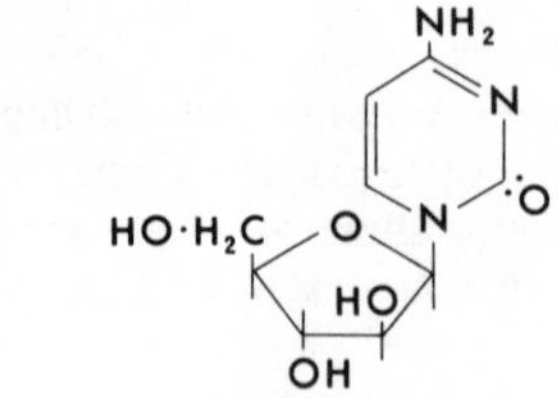

Cytarabine
(*5.24*)

Idoxuridine (5-iododeoxyuridine, IUdR) is converted in cells to idoxuridine-5′-monophosphate which competes with thymidylic acid for incorporation into the DNA of viruses. Clinically it is applied locally to cure the keratitis caused by infection of the eyes with the virus of herpes simplex. This treatment quickly terminates what used to be a long-lasting and painful disease [38]. When virus was exposed to IUdR, most of the thymine was replaced by 5-iodouracil and the DNA did not generate new virus particles [39].

Cytosine arabinoside (*5.24*) (cytarabine) blocks the synthesis of deoxycytosine and hence inhibits formation of DNA. It is so selective that it can be safely given by continuous intravenous infusion (spread over 2–6 days) without seriously affecting the bone-marrow [41]. It is used to cure generalized herpes in man, and is the treatment of choice in myelocytic leukaemia.

Allopurinol (*5.25*), 4-hydroxypyrazolopyrimidine, an analogue of hypoxanthine, is the most effective known remedy for gout. Without being built into a nucleotide, it blocks the oxidation of hypoxanthine by the enzyme xanthine oxidase, and thus reduces the uric acid load of the patient without any side-effects [42].

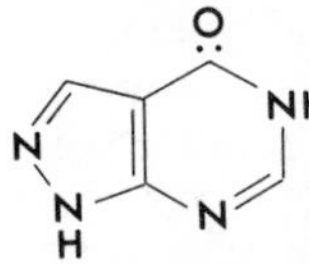

Allopurinol
(*5.25*)

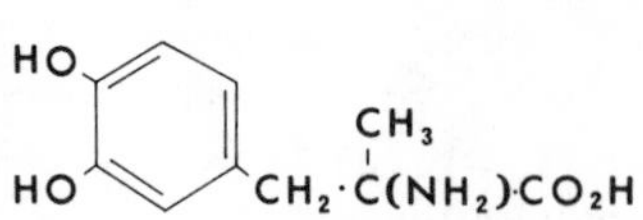

α-Methyl-dopa
(*5.26*)

Some analogues of the neurotransmitters will now be discussed. The biosynthesis of noradrenaline can be decreased by oral administration of methyl-dopa (*5.26*) ('Aldomet'). Chemically it is 2-methyl-3′, 4′-dihydroxyphenylalanine. This is one of the most used of all drugs for lowering very high blood-pressure. In the body, it is decarboxylated and hydroxylated to α-methylnoradrenaline which acts centrally by displacing noradrenaline from its stores [43].

A highly specific agonist, currently bringing relief to sufferers from asthma, is salbutamol (*5.27*). Chemically this is 1-(4-hydroxy-3-hydroxymethylphenyl)-2-*t*-butylaminoethanol. Of the three known receptors for catecholamines, it acts exclusively on the β_2, and hence produces the needed bronchodilation without any of the distressing β_1-activated tachycardia that was a feature of therapy with adrenaline and isoprenaline.

Ephedrine (*5.28*), useful in the treatment of

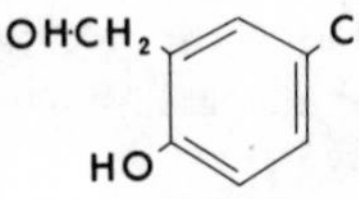

$OHCH_2$ — $CH(OH){\cdot}CH_2{\cdot}NHtBu$, HO

Salbutamol
(*5.27*)

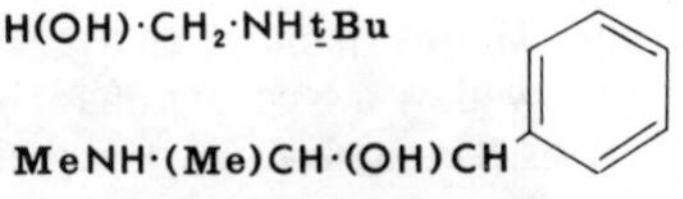

$MeNH{\cdot}(Me)CH{\cdot}(OH)CH$ — (benzene ring)

Ephedrine
(*5.28*)

allergies, acts mainly by releasing noradrenaline from its stores. Unlike this transmitter, ephedrine is active by mouth. The chemically somewhat similar drug phenylephrine ('Neo-synephrine') is a direct agonist.

The antihistamine drugs, which are particularly successful metabolite analogues, were designed by Bovet (in 1937) to contain the essentials of the histamine molecule [44].

Many analogues of acetylcholine (*3.6*) are used in the clinic. Carbachol (*3.5*) was discussed on p. 50. This analogue blocks the acetylcholine-destroying enzyme. Thus its seemingly agonistic action is really produced through antagonism, i.e. it is a selectively toxic agent. Some true agonists of acetylcholine were recognized as long ago as 1914. These are (*a*) nicotine which exerts an acetylcholine-like action at the voluntary muscle-nerve junction, and at ganglionic (nerve-nerve) synapses, whether sympathetic or parasympathetic; and (*b*) muscarine which exerts an acetylcholine-like action at such postganglionic nerve endings as are parasympathetic (mainly involuntary muscle). Thus nicotine and muscarine are more selective than the natural neurotransmitter which has to function at all these sites.

The relative complexity of the structures of nicotine and muscarine has cost many years of guessing how they could mimic acetylcholine. However a pyridine molecule with a simple basic group in the 3-position has the same action as nicotine. Further simplification, followed by adjustment for molecular size and lipophilicity, has produced *N*-pentyltrimethylammonium (*5.29*), an ion as active and selective as nicotine and eight times as active as acetylcholine. Similarly musccarine's specificity is available in as simple a compound as methacoline (*5.30*) (L-acetyl-β-methylcholine).

$Me_3\overset{+}{N}{\cdot}C_5H_{11}$

Pentyltrimethylammonium ion
(*5.29*)

$H_3C{\cdot}\overset{O}{C}{\cdot}O{\cdot}\overset{CH_3}{CH}{\cdot}CH_2{\cdot}\overset{+}{N}Me_3$

Methacholine
(*5.30*)

Many antagonists are known which inhibit the receptors for acetylcholine. For instance tubocurarine (p. 13) acts at the nerve-voluntary muscle junction, hexamethonium (*5.31*) at the parasympathetic ganglia, and atropine at the postganglionic nerve endings. The selectivity of these antagonists has proved very helpful in their clinical applications. As always, antagonists tend to have higher molecular weights than the corresponding agonists. This is clearly seen in the higher homologues of the cation (*5.29*). Whereas the lower members are pure muscarinic agonists, some residual atropine-like antagonism appears when $R = C_6H_{13}$, and when R is made $C_{12}H_{25}$, the action becomes purely atropinic [45].

$Me_3\overset{+}{N}{\cdot}(CH_2)_6{\cdot}\overset{+}{N}Me_3$

Hexamethonium
(*5.31*)

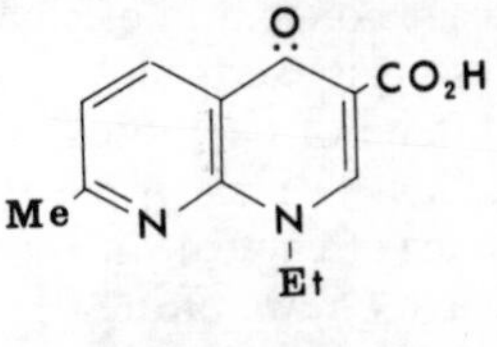

O, CO_2H, Me, N, N, Et

Nalidixic acid
(*5.32*)

Conformational changes. It seems likely that many drugs act on their receptors by inducing a conformational change. Thus aminoacridines manage to find space between the parallel base layers of DNA only by inducing this nucleic acid to unwind slightly (see p. 15).

The first direct evidence that substrates can

move atoms, and even whole groups, at the active sites of enzymes came from an X-ray diffraction analysis of carboxypeptidase with and without a typical substrate (glycyl-tyrosine). The phenyl group of this substrate is absorbed in a deep cavity and this forces the peptide carbonyl oxygen atom (of the substrate) against the zinc atom, which consequently loses a co-ordinated molecule of water. The free carboxyl group of the substrate forms an ionic bond with the enzyme's arginine residue which, forced thus to move 2 Å, disrupts some hydrogen bonds. This disturbance causes the free hydroxy group of the enzyme's tyrosine (residue 248) to rotate through 120° and protonate the nitrogen atom in the susceptible peptide bond which then proceeds to hydrolysis [46].

It is often found that, after repeated application of a drug, a tissue becomes insensitive to it. It has been suggested that this 'tachyphylaxis' reflects a change of the agonist-receptor complex (AR) to a different conformational state (AR′) which, on dissociation, exposes an altered receptor (R′) and this reverts only slowly to the original receptor [47].

5.7 Other utilizable biochemical differences

Insect hormones and pheromones. A high degree of control of physiological action in insects is exerted by hormones chemically different from those found in any other form of life. The selective extermination of insects is being attempted with agonists and antagonists based on two of these hormones: ecdysone and the juvenile hormone. Pheromones, which convey messages between insects, are being studied with similar aims in mind. The slow maturation of desert locusts is accelerated when they crowd together. This suggests that they emit a volatile stimulant; exploration of this phenomenon, along selectively toxic lines, could prevent terrible economic losses in tropical areas.

Inhibitors of protein synthesis. Emetine, an alkaloid, long used in treating the severer cases of amoebic dysentery, inhibits protein synthesis in the mammalian liver, but this is quickly followed by a recovery phase of increased synthesis. The basis of its selectivity lies in the inability of protozoa to achieve a recovery phase [48].

Streptomycin exerts its selective antibacterial action by inhibiting protein synthesis in bacteria without harm to that occurring in mammals. Streptomycin penetrates through the bacterial plasma membrane in a rather complex way that depends on the presence of the bacterial cell wall: much of the selectivity may reside in this process. It then displaces magnesium from ribosomes. It ends on the A-site (of the 30 *S* unit) which it distorts so that neither aminoacyl-t-RNA nor peptidyl-t-RNA can bind to it, and protein synthesis is halted [3].

For some other inhibitors of protein synthesis, see tetracycline (p. 22) and chloramphenicol (p. 60).

Inhibitors of nucleic acid synthesis. Sulphonamides (p. 47), aminoacridines (p. 16), and anti-dihydrofolate reductases (p. 44) have already been dealt with. Nalidixic acid (*5.32*) inhibits the synthesis of DNA in Gram-negative bacteria, but not in Gram-positive bacteria nor in mammals. It is used as a urinary antiseptic, the easily ionized carboxylic acid group guaranteeing excretion by the kidneys [49].

Actinomycin D, which has two pentapeptide side-chains attached to a phenoxazine nucleus (large and flat), intercalates into the DNA template of RNA polymerase, near to a G-C pair. It has proved uniquely valuable in preventing metastases in Wilms's tumour. This solid cancer of the kidney forms a high proportion of all malignant tumours in children [50]. Most other solid tumours require longer treatment, during which the poor selectivity of this drug becomes plain.

2-α-Hydroxybenzylbenzimidazole blocks synthesis of viral RNA in host cells without disturbing synthesis of host RNA [51].

Caffeine, which is consumed daily and in vast quantities by millions of people, without harm, increases the rate of DNA breakdown in *E. coli*, and is highly mutagenic in this bacterium [52]. It is a sobering thought that, under legislation current in many advanced countries, these adverse effects of caffeine could hold up the introduction of tea and coffee, were they not already in common use.

5.8 Conclusion

Differences in the biochemistry of species, as discussed so far, have been mainly qualitative. Yet even where similar metabolic pathways are used by two species, *quantitative* differences become apparent. For example, pathogenic trypanosomes utilize glucose in much the same way as their hosts but about 2000 times faster. Such intense carbohydrate metabolism is all the more vulnerable to attack by drugs because so little of the energy is stored by the parasites.

References

[1] Murray, A.W. (1966), *Biochem. J.*, **100**, 664–674.

[2] Dixon, M. and Webb, E. (1964), *Enzymes*, 2nd Edn, Longmans, London, 782 pp.

[3] Gale, E.F., Cundliffe, E., Reynolds, P.E., Richmond, M.H., and Waring, M.J. (1972), *'The Molecular Basis of Antibiotic Action'* Wiley, London, 456 pp.

[4] Mansour, T.E. and Bueding, E. (1954), *Brit. J. Pharmacol. Chemother.*, **9**, 459–462; Bueding, E. and Fisher, J. (1966), *Biochem. Pharmacol.*, **15**, 1197–1211.

[5] Farber, S. (1952), *Blood*, **7**, 107–112; Zuelzer, W.W. (1964), *Blood*, **24**, 477–494.

[6] Brulé, G., Eckhardt, S.J., Hall, T.C., and Winkler, A. (1973), *Drug Therapy of Cancer*, World Health Organization, Geneva, 163 pp.

[7] Wood, R.C., Ferone, R., and Hitchings, G.H., (1961), *Biochem. Pharmacol.*, **6**, 113–124.

[8] Falco, E.A., Goodwin, L.G., Hitchings, G.H., Rollo, I.M., and Russell, P.B. (1951), *Brit. J. Pharmacol. Chemother.*, **6**, 185–200.

[9] Baker, B.R. and Shapiro, H.S. (1966), *J. Pharm. Sci.*, **55**, 308–317.

[10] Burchall, J., and Hitchings, G.H. (1965), *Molec. Pharmacol.* **1**, 126–136.

[11] Ferone, R., Burchall, J., and Hitchings, G.H. (1969), *Molec. Pharmacol.*, **5**, 49–58.

[12] Jaffe, J.J. and McCormack, J.J. (1967), *Molec. Pharmacol.*, **3**, 359–369.

[13] Roth, B., Falco, E.A., and Hitchings, G.H. (1962), *J. Med. Pharm. Chem.*, **5**, 1103–1123.

[14] Cattell, W., Chamberlain, D., Fry, I., McSherry, M., Broughton, C., and O'Grady, F. (1971), *Brit. Med. J.*, **1**, 377–379.

[15] Richards, W.H.G. (1970), *Adv. Pharmacol. Chemother.*, **8**, 121–147.

[16] Hurly, M. (1959), *Trans. Roy. Soc. Trop. Med. Hyg.*, **53**, 410–412.

[17] Hartley, B.S. and Kilby, B.A. (1952), *Biochem. J.*, **50**, 672–678.

[18] O'Brien, R.D. (1967), *Insecticides, Action, and Metabolism*, Academic Press, N.Y., 332 pp.

[19] Lee, R.M. and Hodsden, M.R. (1963), *Biochem. Pharmacol.*, 12, 1241–1252.

[20] Krueger, H.R. and O'Brien, R.D. (1959), *J. Econ. Entomol.*, **52**, 1063–1067.

[21] Krueger, H.R., O'Brien, R.D., and Dauterman, W.C. (1960), *J. Econ. Entomol.*, **53**, 25–31.

[22] Casida, J.E., McBryde, L., and Niedermeir, R.P. (1962), *J. Agric. Food. Chem.*, **10**, 370–377.

[23] Ashe, H.B. (1964), *Arch. Environ. Health*, **9**, 545–554.

[24] Davies, J.H., Campbell, W.R., and Kearns, C.W. (1970), *Biochem. J.*, **117**, 221–230.

[25] Levine, R.J., Hall, T.C., and Harris, C.A. (1963), *Cancer*, **16**, 269–272.

[26] Levy, D. and Boiron, M. (1969), *Bulletin de Cancer, France*, **56**, 365–374 (in French).

[27] Yamamoto, I. (1970), *Ann. Rev. Entomol.*, **15**, 257–272

[28] Tolmsoff, W.J. (1962), *Phytopathology*, **52**, 755 pp.

[29] Gysin, H. (1962), *Chem. and Indust.*, 1393–1400.

[30] Gibson, F. (1964), *Biochem. J.*, **90**,

256–261
[31] Goodwin, T.W. (ed.) (1965), *Aspects of Insect Biochemistry* (*Biochem. Soc. Symposia*, **25**), London, Academic Press, 107 pp.
[32] Mansour, T.E. (1964), *Adv. Pharmacol.*, **3**, 129–165.
[33] O'Brien, I.G., Cox, G.B., and Gibson, F. (1971), *Biochim. Biophys. Acta*, **237**, 537–549.
[34] Krebs, H.A. (1965), *Endeavour*, **16**, 125–132.
[35] Rogers, E., and 12 others (1960), *J. Amer. Chem. Soc.*, **82**, 2974–2975.
[36] Buchanan, J.M., Flaks, J.G., Hartman, S.C., Levenberg, B., Lukens, L.N., and L. Warren., in *Chemistry and Biology of Purines* (Ciba Symposium) (ed. Wolstenholme, G. and O'Connor, C.), Churchill, London, 233–255).
[37] Reyes, P. and Heidelberger, C. (1965), *Molec. Pharmacol.*, **1**, 14–30.
[38] Kaufman, H.E. (1962), *Proc. Soc. Exper. Biol. Med.*, **109**, 251–252.
[39] Kaplan, A.S. and Ben-Porat, T. (1966), *J. Molec. Biol.*, **19**, 320–332.
[40] Kim, J.H. and Eidinoff, M.L. (1965), *Cancer Res.*, **25**, 698–702.
[41] Chow, A.W., Foerster, J., and Hryniuk, W. (1970), *Antimicrob. Agents Chemother.*, 214–217.
[42] Elion, G.B., Kovensky, A., Hitchings, G.H., Metz, E., and Rundles, R.W. (1966), *Biochem. Pharmacol.*, **15**, 863–869.
[43] Iversen, L.L. (1967), '*The Uptake and Storage of Noradrenaline in Sympathetic Nerves*', University Press, Cambridge, 253 pp.
[44] Bovet, D. (1947), *Rendaconti Istituto superiore di Sanità, Rome*, **10**, 1161–1193.
[45] Paton, W.D.M. (1961), *Proc. Roy. Soc., B*, **154**, 21–69.
[46] Lipscombe, W. (1970), *Accounts of Chemical Research*, (*Amer. Chem. Soc.*), **3**, 81–89.
[47] Katz, B. and Thesleff, S. (1957), *J. Physiol., Lond.*, **138**, 63–80.
[48] Grollman, A.P. (1968), *J. Biol. Chem.*, **243**, 4089–4094.
[49] Goss, W.A., Deitz, W.H. and Cook, T.M. (1965), *J. Bact.*, **89**, 1068–1074.
[50] Farber, S. and Mitus, A. (1968) in '*Actinomycin*' (ed. Waksman, S.), Wiley, N.Y. 231 pp.
[51] Tamm, I., Eggers, H.J., Bablanian, R., Wagner, A.F., and Folkers, K. (1969), *Nature, Lond.* **223**, 785–788.
[52] Grigg, G.W. (1970), *Molec. Gen. Genetics*, **106**, 228–238.

Suggestions for further reading

Albert, A. (1973), *Selective Toxicity*, 5th edn., Chapman and Hall, London, 597 pp.

Covers the territory of this chapter, but in greater depth.

Florkin, M. and Mason, H. (1960–1964), *Comparative Biochemistry* (7 vols) Academic Press, N.Y.

A rich source of reference; however, many advances have taken place since 1964.

Franklin, T.J. and Snow, G.A. (1971), '*Biochemistry of Antimicrobial Action*', Chapman and Hall, London, 152 pp.

Discusses mainly substances of microbial origin such as mycobactins and tetracyclines.

Gale, E.F., *et al*, (as Ref. 3, above).

Experienced investigators discuss the molecular basis of antibiotic action.

6 Favourable differences in cell structure: the third principle of selectivity

It is an everyday observation that the different forms of life differ greatly in their external structures, and this is also the case internally. Plants differ from animals by having photosynthetic devices, and they also have walls around all cells. Animals differ from plants by having nerves, and muscles. The organization of plant and animal cells into a variety of tissues makes a valuable division of labour possible, and further divisions of labour occur at the sub-cellular level. The many, often conflicting, chemical reactions which take place simultaneously in cells require many isolated compartments constructed from membranes of selective permeability. These membranes comprise up to 80 per cent of the dry weight of animals [1]. The electron microscope has revealed, within cells, many kinds of organelle, each with its own function.

If any type of organelle, say the nucleus, is examined closely, it is seen to differ not only from organism to organism, but even between the different tissues of one organism. The outstanding physical differences in the internal organization of mitochondria from human kidney, liver, and brain [2], for example, seem to offer opportunities for selective therapy.

Other opportunities are provided by degrees of differentiation. Most cancer cells are normal, highly specialized cells that have lost some, or much, of their differentiation. As a result, many of them synthesize DNA, and undergo mitosis, at rates much faster than those of the surrounding normal cells. Therapy with hormones, which does not actually injure the malignant cells, restrains this wild mitosis by restoring differentiation [3]. An example is the use of oestrogens in post-menopausal cancer of the breast. For an account of mitosis and the cell cycle, see [4].

The chloroplast, the green photosynthetic organelle in plants, about 0·2 μm in diameter, consists of a double-layered lipoprotein membrane which contains grana and the interconnecting lamellae which carry the chlorophyll molecules. Many herbicides have been found which function by attacking the chloroplasts specifically and without harm to animals. Photosynthesis begins with the absorption of light by quantasomes (aggregates of about 200 molecules of chlorophyll); this brings about the photolysis of water (the Hill reaction). The following herbicides attack at this site: simazine (*5.20*), the phenyl ureas (including the much used diuron; 3, 4-dichlorophenyldimethylurea), acylanilides, and substituted uracils.

Vulnerability to selective toxicants is nowhere more evident than in bacteria whose small size, compared to eukaryotic cells, leaves no space for a nucleus or for even one mitochondrion. In place of a nucleus, the chromosomal DNA is gathered into a strand attached to the plasma membrane; the whole plasma membrane functions as a mitochondrion, breaking down nutrients and storing the energy. Thus the exposed position of the nuclear and mitochondrial functions contrasts sharply with the membrane-protected situation of these functions in the cells of higher organisms. The selectivity of amino-acridines (p. 43), e.g. against bacteria in wounds, exemplifies this structural difference.

O_2N–(benzene ring)–C(H)(OH)–C(H)(NH·CO·$CHCl_2$)–CH_2OH

Chloramphenicol
(6.1)

Whereas ribosomes of higher organisms have a sedimentation coefficient of 80 *S*, those in bacteria sediment at 70 *S*, and are unusually rich in magnesium upon the withdrawal of which they split into two sub-units: one of 50, the other of 30, *S*. The antibiotic chloramphenicol (*6.1*) readily distinguishes between the two ribosomal types for it rapidly halts protein synthesis, but only in bacterial ribosomes. This drug is adsorbed on the 50 *S* unit, where it blocks peptidyl transfer [5]. The nitro-group in chloramphenicol can be replaced by other electron-attracting groups with little loss in activity.

Erythromycin, a macrolide antibiotic which shows little toxicity to man, is strongly bound to the 50 *S* unit of bacterial ribosomes, though not at the site which binds chloramphenicol. It does not bind to mammalian ribosomes.

Another, and very remarkable, structural peculiarity of bacteria is the cell wall, so differently constituted from those found in fungi and the higher plants (cell walls, moreover, are lacking in the animal kingdom). Bacteria are under a high internal osmotic pressure, and when deprived of the restraining network provided by the cell wall, they quite simply burst [6].

This cell wall, which is grossly porous, external to the plasma membrane, is a lipoprotein mosaic, only 4 molecules thick, and highly selective in its permeability. Several antibiotics (oxamycin, penicillin, and bacitracin among others) prevent the synthesis of new cell wall. As soon as bacteria begin to grow in the absence of new wall, they promptly burst and die.

Although the bacterial cell wall has several components, the strong framework that holds it all together is a polymerized carbohydrate with a polypeptide side chain: there is nothing else like this in Nature.

The polypeptide, often called murein, arises by the polymerization of disaccharide building-blocks formed from acetylglucosamine etherified with lactic acid. The most unusual feature is that the carboxylic group of the lactic acid forms the first link in a polypeptide chain, of which a typical example is:

Lactoyl-Ala-Glu-Lys-Ala-Ala

L D L D D

This polypeptide, which differs from species to species, always ends in two D-alanine residues, one of which is lost when cross-linking occurs.

The antiboitic oxamycin (*6.2*) is recognized as a rigidly cross-linked structural analogue of D-alanine (*6.3*). It has been shown to prevent the incorporation of D-alanyl-D-alanine into the polypeptide when a new bacterial wall is being formed [7].

H_2N·C(H)–CH_2–O–NH–C(:O) (ring)

Oxamycin
(6.2)

H_2N·C(H)(–CH_3)–C(:O)OH

D-Alanine
(6.3)

Penicillin acts in a similar, but much more effective way. As pointed out by Tipper and Strominger [8], it is a structural analogue of D, D-dipeptides. This can be seen by comparing formulae (*6.4*) and (*6.5*), which are two-dimensional projections of molecular models. If N' in the dipeptide is superimposed on N' of penicillin, then N'' falls on N'', and C' on C'. These authors found that the penicillins irreversibly acylate a transpeptidase, the normal function which is to receive both D-alanine residues of the side-chains of two of the cell-wall forming saccharopeptide units, and (by uniting these side-chains) to commence the series of polymerizations that lead to new cell wall. Thus the penicillins (and the cephalosporins, too), by blocking this enzyme, prevent linkage of the units, with the result that the bacterium dies.

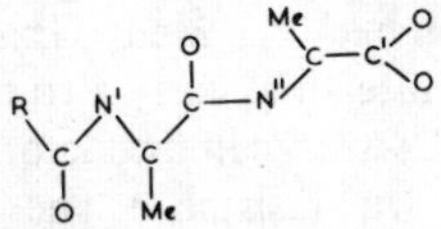

(Acyl)-D-alanyl-D-Alanine (*6.4*) 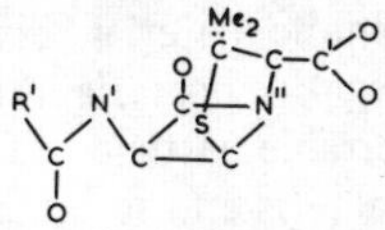   The penicillins (*6.5*)

Specific agents for interfering with the synthesis of fungal cell wall [9], and with the structure of fungal plasma membrane (using polyene antibiotics) are in daily use.

A glance at the electron micrographs in a contemporary book on viruses [10], will show convincingly that many of them have an immensely more complex structure than was formerly thought. This fact, plus their intrinsic vulnerability (e.g. from lack of an energy-producing metabolism) should give opportunities for direct, selective attack. To date, all clinically used antiviral drugs act in other ways, e.g. by preventing penetration of the host's cells or by halting assembly of replicated components in those cells.

The vesicles in which neurotransmitters are stored provide a further example of structural selectivity. The principal action of DDT is to overstimulate the sensory axons of insects. It seems to do this by forming an electron-transfer complex with insect (but not mammalian) acetylcholine-containing vesicles, and this leads to their rupture, causing convulsion followed by exhaustion [11].

References

[1] O'Brien, J.S. (1967), *J. Theoret. Biol.*, **15**, 307–324.

[2] Lehninger, A. (1971), *The Mitochondria*, 2nd edn., W.A. Benjamin, N.Y., 263 pp.

[3] Brulé, G., Eckhardt, S.J., Hall, T.C., and Winkler, A. (1973), *Drug Therapy of Cancer*, World Health Organization, Geneva, 163 pp.

[4] Shall, S. (1975), *The Cell Cycle*, Chapman and Hall, London, 64 pp.

[5] Gale, E.F., *et al.* (see ref. 3, Chapter 5).

[6] Lederberg, J. (1957), *J. Bact.*, **73**, 144.

[7] Strominger, J.L., Threnn, R.H., and Scott, S.S., (1959), *J. Amer. Chem. Soc.*, **81**, 3803–3804.

[8] Tipper, D.J. and Strominger, J.L. (1965) *Proc. Nat. Acad. Sci., U.S.*, **54**, 1133–1141.

[9] Maeda, T., Abe, H., Kakiki, K., and Misato, T., *Agric. Biol. Chem., Japan*, **34**, 700–709.

[10] Fenner, F. and White, D. (1970), *Medical Virology*, Academic Press, N.Y., 390 pp.

[11] Holan, G. (1971), *Nature, Lond.*, **232**, 644–647.

Suggestions for further reading

Bourne, G. (1970), *Division of Labour in Cells*, 2nd. edn, Academic Press, N.Y.

Albert, A. (1973), *Selective Toxicity*, 5th edn., Chapman and Hall, London, 597 pp.
Amplifies subjects discussed in this chapter.

Ashworth, J.M. (1973), *Cell Differentiation*, Chapman and Hall, London, 64 pp.
Introduces modern ideas of how cell differentiation occurs.

Porter, K. and Bonneville, M. (1969), *Fine Structure of Cells and Tissues*, 3rd. edn., Lea and Febiger, Philadelphia.
Descriptive atlas of the structure of cells and tissues.

7 Acquired resistance to drugs: the loss of selectivity

A drug designer may build the highest degree of selectivity into his agents, only to see it lost through the development of resistance in his target organism. Drug resistance can arise in at least five distinct ways.

Type 1 resistance takes place by exclusion of the agent. Thus *Staphylococcus aureus* becomes resistant to tetracycline by abolishing the drug-accumulating mechanism described on p. 22; the drug's target remains as sensitive as ever, but the drug no longer reaches it.

Type 2 resistance is accomplished by the organism bringing a destructive enzyme into play. This arises by the selection of a favourable mutant, or by induction. Penicillin-resistant strains of *S. aureus*, isolated from patients, contain penicillinase which hydrolyses the drug to useless penicilloic acid.

By this mechanism, over 200 species of insects have developed ability to withstand one or more major insecticide(s) formerly effective against them.

Type 3 resistance is brought about by the organism withdrawing a necessary drug-modifying enzyme. Many purine and pyrimidine analogues do not act against cancer until intracellular enzymes metabolize them to ribonucleotides (see p. 54). Mouse leukaemic cells become resistant to 6-mercaptopurine in this way.

Type 4 resistance takes the form of secretion, by the organism, of an excessive amount of the substance to which the drug is a metabolite antagonist. Thus pneumococci becomes resistant to antibacterial sulphonamides by forming extra quantities of *p*-aminobenzoic acid.

Type 5 resistance is due to the exchange of genetic material between bacteria. Many Gram-negative bacteria in the intestines carry a transferrable particle of DNA which can invade other bacteria and so transfer drug-resistance. The bacteria in question are those that cause dysentery, cholera, typhoid and plague. In this type, the drug (e.g. tetracycline or chloramphenicol) is chemically altered to an inert substance.

When one pauses to consider how widely selectively toxic agents have been used, it is surprising that more types and cases of resistance have not been encountered. Clearly the versitility of organisms, in this respect, is limited. It is true that the resistance of staphylococci, pneumococci, and gonococci to penicillin is clinically most disturbing, but in no other type of infection has an embarrassing degree of penicillin-resistance arisen after 30 years of intensive use. Although staphylococci develop a high degree of resistance, streptococci do not. In all the long years that syphilis was treated with arsenicals, resistant spirochaetes were never found, not has this organism developed resistance to penicillin.

For further reading, see Albert, A. (1973), *Selective Toxicity*, 5th edn., Chapman and Hall, London, 597.

Index